TITANIC'S UNLUCKY SEVEN

TITANIC'S UNLUCKY SEVEN

THE STORY OF THE ILL-FATED LINER'S OFFICERS

JAMES W. BANCROFT

FRONTLINE
BOOKS

First published in Great Britain in 2024
by Frontline Books
An imprint of
Pen & Sword Books Ltd
Yorkshire - Philadelphia

Copyright © James Bancroft

ISBN 9 781 03610 251 7

The right of James Bancroft to be identified as Author of this work has been asserted by him in accordance with the Copyright, Designs and Patents Act 1988.

A CIP catalogue record for this book is available from the
British Library

All rights reserved. No part of this book may be reproduced or transmitted in any form or by any means, electronic or mechanical including photocopying, recording or by any information storage and retrieval system, without permission from the Publisher in writing.

Typeset by Lapiz Digital
Printed and bound in the UK by CPI Group (UK) Ltd,
Croydon, CR0 4YY.

Printed on paper from a sustainable source by
CPI Group (UK) Ltd, Croydon, CR0 4YY

Pen & Sword Books Limited incorporates the imprints of Archaeology, Atlas, Aviation, Battleground, Digital, Discovery, Family History, Fiction, History, Local, Local History, Maritime, Military, Military Classics, Politics, Select, Transport, True Crime, Air World, Claymore Press, Frontline Publishing, Leo Cooper, Remember When, Seaforth Publishing, The Praetorian Press, Wharncliffe Books, Wharncliffe Local History, Wharncliffe Transport, Wharncliffe True Crime and White Owl.

For a complete list of Pen & Sword titles please contact
PEN & SWORD BOOKS LTD
47 Church Street, Barnsley, South Yorkshire, S70 2AS, England
E-mail: enquiries@pen-and-sword.co.uk
Website: www.pen-and-sword.co.uk
or
PEN & SWORD BOOKS
1950 Lawrence Rd, Havertown, PA 19083, USA
E-mail: uspen-and-sword@casematepublishers.com

CONTENTS

Introduction ix

Chapter 1: In the Beginning. 1
Chapter 2: Report for Duty 5
Chapter 3: 'All Aboard!'. 13
Chapter 4: 'We've Hit Something!'. 21
Chapter 5: 'Man the Lifeboats!'. 27
Chapter 6: Peril on the Sea. 39
Chapter 7: '*Titanic* Disaster: Great Loss of Life' 45
Chapter 8: Concise Biographical Tributes 55
Chapter 9: Did Captain Smith Survive the Sinking? 79
Chapter 10: The British *Titanic* Inquiry 89

Bibliography and Research Sources 223
Index 229

Chief Officer Henry Tingle Wilde
First Officer William McMaster Murdoch
Sixth Officer James Paul Moody

'I don't think I'll ever be sure again, about anything.'
Kenneth More as Second Officer Charles Herbert Lightoller in
A Night to Remember

INTRODUCTION

The RMS *Titanic* disaster that occurred in the North Atlantic Ocean on the night of 14/15 April 1912, when the vessel collided with an iceberg and sank, is one of history's most enduring catastrophic human tragedies, which resulted in a terrible sacrifice of life. The incident still captures the interest and imagination of people all over the world, and a study of the individuals involved provides a time warp, which reflects the late Victorian and early Edwardian era. The people on board were proud to travel on the maiden voyage of the ship that all the world was talking about, but as the officers reported for duty in Belfast and Southampton, they did not know it was destined to be *Titanic*'s only voyage.

After *Titanic* hit the iceberg, there were only seven officers to take charge of hundreds of people who minutes earlier were relaxing in the company of friends with not a care in the world, or snuggled warm and cosy in their beds; now they were out in the cold air of the mid-Atlantic Ocean, in great peril of their lives. How does anyone cope with that situation, especially with some element of sleep deprivation? As Fifth Officer Harold Lowe put it when referring to the four hours on and four hours off disturbed rest pattern: 'When we sleep we die.'

While some passengers stated that the officers were harsh and uncompromising – and it has to be said that perhaps the dreadful events warranted them to be – several were of the opinion that each one of them behaved with great courage and discipline in a situation beyond anything they had previously experienced. If the officers, particularly Charles Lightoller and Harold Lowe, had not taken charge in the way they did, it is unlikely that the evacuation would have succeeded and there would have been no survivors. Three officers lost their lives during the tragic events, and after studying the subject of gallantry medals for five decades, it is my opinion that it was only the desire of the authorities to try to cover up the failings of the incident that prevented a number of gallantry medals, such as the Albert Medal, to be awarded to some of the officers and passengers. None of the four surviving officers were ever given command of a ship after the incident, when their experience should have entitled at least one or two of them to have been appointed to a senior command.

Henry Tingle Wilde was scheduled to sail with *Titanic*'s sister ship, RMS *Olympic*, but he was switched to *Titanic* as the chief officer. He reported for duty on the very day the ship departed Southampton. This move meant a reshuffle of the officers and, as only seven officers were deemed necessary – an insufficient number as it turned out – Second Officer David Blair was removed from the crew list and sent ashore. He was certainly the luckiest of all, as Wilde went down with the ship.

Of the many questions asked about the sinking of *Titanic* is the fate of Captain Edward Smith. The circumstances of what are said to have been his final moments are confused, and have never been proven, and his body was never recovered. Therefore, it has been assumed that he went down with his ship. However, there are suggestions that he actually survived the sinking, and was seen and spoken to in Baltimore by a 'perfectly sane' man who had sailed with him, and who had known him for much of his life. Certainly, Captain Smith had good reason to disappear into obscurity, and could it really have been a case of misidentification of both sight and voice on two occasions?

By describing the scene through the eyes of just seven men, I have attempted to allow the reader to see the events as they unfolded in a concise way; with informative biographical tributes to give some background of who they were. In addition to this, I have re-checked the known official documentation about them through Census Returns, Ancestry.com, FindMyPast, FamilySearch, and wherever possible I have obtained birth, marriage and death certificates. I then cross-referenced all the information with my JWB Historical Archive, which I have compiled over five decades; and revised where necessary. The result provides the most information about the *Titanic* officers than has ever before been compiled in one mainstream publication.

***

In the weeks before *Titanic* set off on her maiden voyage news began to reach Britain of a human tragedy of classic proportions. Captain Robert Falcon Scott (1868–1912) and his 'Terra Nova' Antarctic expedition had slogged their way to the South Pole, only to find that they had been beaten to their goal by the Norwegians led by Roald Amundsen (1872–1928). After suffering unthinkable hardships, Captain Scott and four other members of his party had perished during their imperilled return journey.

Realising he was holding them back, one of their number, Captain Lawrence Edward Grace Oates (1880–1912), of the Inniskilling Dragoons, had sacrificed his own life in the hope that his comrades would survive by walking out of their tent into a raging blizzard, with the words 'I am just going outside, and I may be some time.' He never returned.

Little did the officers of *Titanic* know that very soon they too would find themselves in a horrific situation when each one of them would have to be, as Captain Scott said of Captain Oates: 'A very gallant gentleman'.

James W. Bancroft,
2024

Chapter 1

IN THE BEGINNING

The White Star Line was established in 1845 by two Liverpool men, John Pilkington (1820–1890), who was the son of the Pilkington Glass Company's founder, Christopher Pilkington, and who exhibited at the Great Exhibition at The Crystal Palace in Hyde Park in 1851; and Henry Threlfall Wilson (1825–1869). They began the business of shipbrokers at Prince's Building, 26 North John Street in Liverpool, from where they initially leased and chartered ships to operate packet sailing to the east coast of the United States. They started to use the name White Star Line of Boston Packets in 1849, but shortened it to White Star Line after they started sending ships to Australia. Business increased after the discovery of gold in Australia in 1851, and this enabled them to actually purchase the vessels they used.

The company's bank failed in 1867 and with massive debts it was forced into bankruptcy. However, the following year Thomas Henry Ismay (1837–1899) purchased the house flag, trade name and goodwill of the bankrupt company, with the intention of operating large steamships on the North Atlantic service between Liverpool and New York, and he established the company's headquarters at Albion House near Pier Head in Liverpool (now known as 30 James Street).

During a game of billiards, Ismay was approached by the shipbuilder, Gustav Wilhelm Wolff (1834–1914), and his uncle, Gustav Christian Schwabe (1813–1897), who was a prominent Liverpool merchant, who offered to finance the new line. A partnership with Harland and Wolff was established in summer 1869, the agreement being that Harland and Wolff would build ships exclusively for the White Star Line. On the death of Ismay in 1899, he was succeeded as the chairman and managing director of the White Star Line by his son, Crosby-born and Harrow-educated, Joseph Bruce Ismay, known as Bruce (1862–1937). In 1907 Bruce Ismay met Lord

William Pirrie (1847–1924) of Harland and Wolff, to discuss White Star Line's answer to Cunard Line's recently unveiled RMS *Lusitania* and RMS *Mauretania*. They came up with the idea of building three 'Olympic Class' ships at Harland and Wolff that were planned to be marvels of engineering.

Under the heading 'White Star Line', several Australian newspapers reported on 27 January 1912:

> A sign of the continued growth in the trade in Australasia is the request made for additional berthing accommodation to Tilbury Dock, London, on the part of the White Star Line. Messrs Ismay, Imrie and Company, has applied to the Port Authority for the use of a berth on the south side of the main dock for the vessels of their Australian Line, which will include the largest vessels coming to the port. To meet their requirements it will be necessary to extend the dock wall, create a shed, and provide trains and a railway track, at an estimated cost of £52,000. The Port Authority has decided to let the berth in question to Messrs Ismay, Miry and Company, on the usual conditions, and to incur the necessary expenditure for carrying out the alterations and improvements.

***

The Harland and Wolff shipbuilding yard had been established in Belfast in 1861. In his capacity as a journalist, Abraham 'Bram' Stoker (1847–1912), famous for his gothic novel *Dracula* visited the shipyard, and afterwards stated:

> Less than fifty years ago the firm of Harland and Wolff was a small, un-ambitious concern. It was only when the manager, Mr – afterwards Sir Edward Harland [1831–1895] – acquired possession that expansive power began to manifest itself. It was not; however, until Lord Pirrie took command that full development was reached. For close on forty years he has been connected with the firm, first as partner, latterly as head.
>
> In this shipyard it is possible to follow the whole process of construction, from the reception of the raw material – in itself a big work – to the departure of the registered ship. The Shipyard [sic] proper is surrounded on three sides by water; to the south the Abercorn Basin, to the north and west the River Lagan. On the south end are five slips – always occupied – and on the north four, and no sooner is a vessel launched than preparations begin for laying the keel of another.

Bram Stoker died less than a week after the *Titanic* disaster.

***

The three ships planned for the Olympic Class would focus on comfort and luxury over speed, with lavishly decorated rooms, and passenger facilities of the highest standard. They would be fitted with special gears that would reduce vibration from being transmitted to the rest of the ship and the passengers. Simply by pulling a level on the bridge it was possible to close six compartments within the vessel so that if water was to enter in some way it would be contained within these compartments.

The first of these 'unsinkable' ships was named *Olympic*; ordered in 1907, and construction began on 16 December 1908. *Olympic* was launched on 20 October 1910, completed on 31 May 1911 (the same day that *Titanic* was launched), and she began her maiden voyage to New York, via Cherbourg on the western coast of France, and Queenstown (now Cobh) on the southern coast of Ireland, on 14 June 1911, reaching New York a week later.

The outward trip had gone as planned, but on the return voyage, Washington Atlee Burpee (1858–1915), the well-known cultivator and founder of what is now known as Burpee Seeds, realised that he had left his spectacles back at his office at Fordhook Farm in Pennsylvania, and sent a wireless to ask for them to be sent on to his London office. Tommy Sopwith (1888–1989), the famous British pioneering aviator, heard of Burpee's plight and volunteered to fly out to the ship in his plane to attempt to drop the glasses onto the deck. However, he missed his target and they landed in the sea and were lost.

During her fifth voyage on 20 September 1911, *Olympic*, with Captain Edward Smith and Officer William Murdoch on board, had a serious collision with the British cruiser HMS *Hawke*. *Olympic* was blamed for the collision, and newspapers on 19 December 1911 reported '*Olympic-Hawke* Collision – *Olympic*'s Pilot to Blame':

> In giving his reserved decision today regarding the SS [*sic*] *Olympic* and HMS *Hawke* collision case, the president of the Admiralty Court, Lord Justice Evans, said that the blame lay with the *Olympic*. The pilot of the *Olympic* took too wide a sweep round the West Bramble buoy, though having the cruiser *Hawke* close the *Olympic*'s starboard [right] quarter, the pilot should have made way.
>
> The main issues in the action were whether *Hawke*, while coming up astern on the starboard quarter of *Olympic* overhauled the latter vessel, and while *Olympic* kept her course the *Hawke* suddenly altered her course towards *Olympic*, and although the helm of *Olympic* was put hard aport, to try and throw her quarter clear, the *Hawke* truck the starboard quarter of the *Olympic* a heavy blow, causing her serious damage. The *Olympic* further alleged that the helm of the *Hawke* was improperly starboarded, her engines were not eased, stopped, or reversed in due time, and that she failed to indicate her manoeuvres by appropriate whistle signals.

The defendant, the commander of the *Hawke*, pleaded that those in charge of the *Olympic* were negligent – (a) in coming too close to the *Hawke* and proceeding at a speed which was excessive; (b) in entering the Solent Channel at an improper time, and in an improper manner; (c) in taking too wide a sweep round the West Bramble buoy; (d) in not starboarding sufficiently or in due time; (e) in not easing, stopping, or reversing her engines in due time; (f) in not indicating her manoeuvres by the proper sound signals, and as an alternative plea, that if the *Hawke* was the overtaking vessel the *Olympic* did not keep her course and speed as required by the regulations.

White Star Line appealed against the decision but it was dismissed.

The second ship in the class, to be named *Titanic*, was ordered on 17 September 1908, and was laid down for construction on 31 March 1909. *Titanic* was launched on 31 May 1911. A couple of weeks before his seventh birthday, a Belfast boy named William MacQuitty (1905–2004) watched the launch, and in 1958 he became noted for his production of A *Night to Remember*, a film which recreates the story of the *Titanic* disaster.

On 3 February 1912, *Titanic* was dry docked, the move taking a little more than two hours. She would remain in the dry dock for two weeks, while furnishing of her interiors continued; the primary reason for dry docking was to fit her three giant propellers and a final preparation of the hull for the sea; this involved cleaning the hull below the waterline and applying red anti-fouling paint.

However, on 6 March 1912, *Titanic* was moved out of the dry dock to allow for *Olympic* to have a propeller blade fixed, which had become damaged during the collision with *Hawke*. Consequently, *Titanic* was not completed until 2 April 1912, and was fatefully re-scheduled to begin her maiden voyage on Wednesday, 10 April 1912.

At the time of *Titanic's* launch, *The Daily Telegraph* published some particulars about the ship, which, together with her sister ship *Olympic* had cost £3 million:

Tons, 45,000; length between perpendiculars. 882 feet; breadth, 92 feet 6 inches; depth 62 feet; speed, 21 knots; built 1911. The largest beam in the *Titanic* weighs more than four tons, and measures 92 feet. The longest steel plates are 36 feet, and there are two and a half million rivets in the ship. The *Titanic* will have, in addition to dining saloons, lounges, drawing rooms and smoking rooms, several restaurants and veranda cafe. She is to have a splendidly equipped Turkish bath, a swimming bath, and a full-sized racquets court. Passenger accommodation is planned for 750 first-class, 515 second-class, and 1,100 third-class, and the crew will number 860.

Chapter 2

'REPORT FOR DUTY'

The first captain of *Titanic* was in fact Herbert Haddock (1861–1946). He was aged 51, and having signed up as the master of *Titanic* at Liverpool on 25 March 1912, he travelled to Belfast to oversee the crew assembling there for the ship's delivery trip to Southampton. He was to be replaced by Captain Smith when they reached Southampton.

Most of the men who were assigned to the positions of officers on *Titanic* received a telegram directing them to report to White Star Line's main office at Albion House in Liverpool, at 9.00 am on the morning of 26 March 1912. There they collected their tickets to travel to the Harland and Wolff shipyard in Belfast, where they arrived around noon on the following day, and reported to First Officer William Murdoch. They all saw *Titanic* for the first time on that day, 27 March 1912. They must have been highly impressed with the colossal ship, as Chief Officer Henry Wilde wrote from Queenstown to his daughter, Jennie, on 11 April 1912, describing *Titanic* as,

> ... a very fine ship and an improvement on the *Olympic* in many ways. I would like you to see her ...

However, they were not so much impressed when they found that they were bunked in rooms, 'no bigger than a broom cupboard.' Their duties in general were quite simply to do anything they were told to do.

Not counting the captain, Chief Officer Lieutenant William McMaster Murdoch was the oldest of the officers at the age of 39. He was a married Scotsman, who had joined White Star Line in 1900 and the Royal Naval Reserve in 1902. He was the only one of the *Titanic* officers to pass all the Board of Trade examinations at the first attempt. For his most recent assignment he had served aboard *Olympic* since May 1911, and he was on that ship when she collided with *Hawke*. Two other *Titanic* officers had served with him on *Olympic*, Captain Smith and Chief Officer Wilde, and they would join him at Southampton.

Charlotte Caroline Collyer (1881–1916), a surviving passenger, said of First Officer Murdoch:

> He was a masterful man, astoundingly brave and cool ... and thought him a bull-dog of a man who would not be afraid of anything.

Official documentation described Murdoch as being 5 feet 9 inches tall, with a fair complexion, hazel/brown eyes, and brown hair.

Second Officer Sub-Lieutenant Charles Herbert Lightoller, known as 'Lights', was a single man, aged 38, who had joined White Star Line in 1900; and the Royal Naval Reserve the following year. He was a member of a family who were prominent in the Lancashire cotton trade, and his house was one of the first to have electric lighting in his home town of Chorley. He had already been involved in three serious maritime incidents, including being shipwrecked on a remote island, and he had been caught up in a cyclone and nearly died of malaria. He had suffered several domestic tragedies in his life, which had resulted in the death of his mother and three siblings. He served under Captain Smith on RMS *Majestic*, then as third officer on RMS *Oceanic II*, the flagship of the White Star Line. He returned to *Majestic* as first mate, and then back to *Oceanic II* in the same position. Third Officer Herbert Pitman and Sixth Officer James Moody had also served with him on *Oceanic II*. A certain Captain Peter Pryal had served on *Majestic* several years previously.

Second Officer David 'Davy' Blair was the 37-year-old son of a Scotsman who was a British Army colonel. He was born on the Isle of Wight, and had joined the White Star Line in 1902 and the Royal Naval Reserve in 1904. He had married near Dundee in 1905, and his last ship was RMS *Teutonic*. He would be replaced by Chief Officer Wilde when they reached Southampton.

Third Officer Mr Herbert John Pitman, was the 35-year-old unmarried son of a West Country farmer, who had died when he was very young and his mother had remarried. After serving on several ships with other companies, he had moved to the White Star Line in 1906, and he had recently served as second officer on *Oceanic II*. He is known to have suffered from seasickness. Descriptions of him were varied over the years, but at the time he was on *Titanic* he was about 5 feet 10 inches tall, and weighed about 185 pounds. He became a member of the Hatfield Abbey Lodge of Freemasons in 1909, remaining so until he died, and he was also keen on stamp collecting all his life. He was the only *Titanic* officer to have been a native of a county in the south of Britain.

Fourth Officer Sub-Lieutenant Joseph Groves Boxhall, aged 28, was an unmarried man, who joined White Star Line in September 1907, and had been confirmed as sub-lieutenant in the Royal Naval Reserve as recently as 1 October 1911. He had served on the liners *Oceanic II* and SS *Arabic*. He was born in Hull, and it was three days after his twenty-eighth birthday when he travelled to Belfast to board *Titanic*. Like Herbert Pitman he suffered from seasickness. Just after the First World War he was described as being 5 feet 8 inches tall and weighed 154 pounds, although four years later he had shrunk an inch and lost 11 pounds.

Fifth Officer Sub-Lieutenant Harold Godfrey Lowe, was a single man from north-west Wales. He too had suffered a family tragedy, when, at the age of 13, his older brother was drowned in a boating accident, and he had almost lost his own life in a similar way in the following year. He had joined White Star Line only fifteen months prior to boarding *Titanic*, and he had recently served as third officer on SS *Belgic*. He was aged 29, but was said to be 'Very young, and boyish-looking.' He was 5 feet 8 inches tall, with a dark complexion, brown eyes and brown hair, and he had the letters HGL in a heart tattooed on his right forearm. Fireman Thomas 'Tom' Threlfall (1867–1934) considered him to be, 'A gentleman and a Britisher', and after the disaster a female passenger described him as, 'A leader with a cool head, desperate courage, and a knowledge of the sea.' He stated that he had, 'experience with the different classes of ships afloat, from the schooner to the square-rigged sailing vessel, and from that to steamships of all sizes', but he had never worked with any of the other officers before, and this was going to be his first transatlantic crossing.

Sixth Officer Sub-Lieutenant James Paul Moody, was a tall 24-year-old single man from Scarborough, who was the youngest of the officers. He came from a family that was prominent in the public service of Grimsby and Scarborough. He had joined White Star Line in August 1911, and had served aboard *Oceanic II*. He was feeling somewhat peeved, and he had been reluctant to accept the *Titanic* assignment. Having endured a harsh winter, he was hoping to be allowed to take some time off, but his request for leave had not been granted. He was described as being 5 feet 11 inches tall, with a fair complexion, light brown hair and blue eyes.

Captain Herbert 'Bert' James Haddock, known as 'Daddy', was a highly experienced 51-year-old from Rugby. He was married with four children. He served his apprenticeship aboard HMS *Conway* in the River Mersey, where he played for the football team. He joined the Royal Naval Reserve in 1877, and White Star Line in 1888; his main ships being RMS *Cedric* and RMS *Oceanic*. He was made Companion of the Bath (CB) as one of the awards to celebrate the coronation of King Edward VII in 1901. On 31 March 1912, Captain Haddock was informed that he was to be relieved by Captain Smith as master of *Titanic*, and after travelling back to Southampton he took command of Smith's previous ship, *Olympic*, which set sail on its tenth voyage to New York on 3 April 1912.

The sea trials for *Titanic* were arranged to take place on 1 April 1912 but they were postponed because of high winds, and were conducted on 2 April instead. Second Officer Lightoller said they took place in Belfast Lough, where the weather was clear and good, but for a light breeze; but Third Officer Pitman testified that some of the trials took place in the open sea (Irish Sea). All seven officers present in Belfast were aboard, along with Thomas Andrews Jr (1873–1912), the ship's main designer, and about thirty other members of the crew.

The trials consisted of turning the ship in circles to see what space she will turn under certain helms, with the engines at various speeds, and while adjusting the compass. This took about five hours, and then the vessel was driven full speed ahead for two hours out along the 15-mile-long lough, and two more hours back, the fastest speed attained was apparently 23 knots. The final test was full speed astern, to see at what distance the ship would stop with the engines at full speed when going backwards.

At that time First Officer Murdoch instructed his fellow officers to check the lifeboats and their equipment, and during the United States Senate inquiry, Fifth Officer Lowe stated:

> Mr Moody and myself and Mr Pitman and Mr Boxhall took the port [life]boat – that is, I took the starboard, and they took the port [left], and we overhauled them; that is to say, we counted the oars, the rowlocks, or the hole pins, whichever you like to call them, and saw there was a mast and sail, rigging, gear, and everything else that fitted in the [life] boat, and plugs, and also that the biscuit tank was all right, and that there were two breakers in the [life]boat, two bailers, two plugs, and the steering rowlock, that is, the rowlock for the oar that you ship aft where there is a heavy sea running, because you can not steer by rudder when there is a heavy sea running, and you put an oar over and you have greater command over an oar and can put more power on it.
>
> Everything was absolutely correct with the exception of one dipper. A dipper is a long thin can about that length [indicating] and about that diameter [indicating] – an inch and a quarter diameter – and you drop it down into the water breaker and draw the water. That was the only thing that was short out of our [life]boats, and our [life]boats were respectively, numbers 1, 3, 5, 7, 9, 11, 13, and 15, from 1 to 15 – odd numbers. Then the even numbers were on the other side, that is, on the port side of the ship.
>
> We found 14 oars, and anyhow, a set and a half of oars on one set of rowlocks. That is, if there were six rowlocks, there were nine oars in case

of emergency. That is, if an oar got broke there was another extra oar to replace that oar, and there were three spare ones – that is, one and a half sets. If there were 12 oars in one boat, it was fully equipped. There would be 18 oars altogether – six extras – and dippers and everything else. Everything was absolutely correct; I will swear to that. There is a compass, a light, and oil to burn for eight hours; biscuits and water. That is all I can think of at present.

***

At 61 years of age, Captain Edward John Smith, a married man from Stoke-on-Trent in Staffordshire, had more than thirty years of experience working with White Star Line.

He had been involved in five maritime accidents. Several months before joining *Titanic*, he had spoken about his bad luck at sea with a businessman named J.P. Grant, stating that he felt that he had been jinxed, and said that he would resign if he had another accident in a liner.

However, at the time of the disaster Chief Officer Robinson, of the steamer HMS *Euryalus*, stated:

> I was formerly an officer under the *Titanic* commander, and I know him. Captain Smith was a cool, calm, self-contained man. In the stress of danger the more level-headed he would become.

First Officer John Simpson of White Star Line steamer SS *Afric* stated:

> A better commander you could not find. Eight years ago I served under Captain Smith, and in my experience and that of many others he was a fine man in every sense of the word. Standing over six feet high, with a white beard that gave him quite a patriarchal appearance, Captain Smith presented a striking figure. Of a kindly, cheerful disposition, the commander was a great favourite with all who worked under him.

At over 6 feet in height, Henry Wilde was the tallest officer on board *Titanic*. A Liverpudlian, who was orphaned at the age of 9, and he had suffered a terrible tragedy when his baby twins died soon after birth, and his wife died on Christmas Eve 1910. He joined White Star Line in 1897 and the Royal Naval Reserve in 1902. He was described as being 6 feet 1 inch tall, with a dark complexion, blue eyes and dark brown hair. Second Officer Lightoller said of him: 'He was a fine fellow and one for whom I had the greatest admiration.' He had served with Captain Smith on *Olympic* and was on board her when she collided with *Hawke*. He had expected to be transferred to SS *Cymric*, but that vessel was laid-up because of the coal strike. On 7 April 1912 he had written to his family,

> I am on '*Titanic*' but I am not sure I am sailing on her yet. If I go on this ship we sail on Wednesday and will be back in 17 days.

Two days later he received a telegram confirming him as chief officer on *Titanic*.

Having served as second officer on *Majestic* since 5 February 1912, David Blair was assigned to *Titanic* on 26 March 1912. In his 1936 book Titanic *and Other Ships* Charles Lightoller recorded:

> Unfortunately whilst in Southampton we had a re-shuffle amongst the senior officers, owing to the *Olympic* being laid up, the ruling lights of the White Star Line thought it would be a good plan to send the chief officer of the *Olympic*, just for one voyage, as chief officer of the *Titanic*, to help, with his experience of her sister ship. This doubtful policy threw both Murdoch and me out of our stride; and apart from the disappointment of having to step back in our rank, caused quite a little confusion. Murdoch from chief, took over my duties as First, I stepped back on Blair's toes, as Second, and picked up the many threads of his job, whilst he – luckily for him as it turned out – was left behind. The other officers remained the same. However, a couple of days in Southampton saw each of us settled in our new positions and familiar with our duties.

Blair was considered to be too senior to take the position of third officer and was tasked to another ship. He himself wrote to his sister-in-law at Broughty Ferry:

> I am afraid I shall have to step out to make room for the chief officer on *Olympic*. This is a magnificent ship. I feel very disappointed I am not to make her first voyage. I hope eventually to get back to this ship.

Second Officer Blair left the ship on 9 April 1912, and as it was due to sail on the following day he would have been rushing about getting ready to disembark. In his rush he forgot to hand over the key to the locker that contained the binoculars for the look-outs in the crow's nest. It was to be hoped they did not get into any difficulty that would require them. He returned to *Majestic* on the following day.

His daughter, Elizabeth Nancy, wrote in 1958:

> In the rush to pack his belongings and get off before she sailed, he came away with a key in his pocket and there was no opportunity to return it.

The chief, first and second officers were considered to be 'senior', the other four 'junior'; the order of seniority among the seven officers below Captain Edward Smith became:

Chief Officer Henry Tingle Wilde
First Officer William McMaster Murdoch
Second Officer Charles Herbert Lightoller
Third Officer Herbert John Pitman
Fourth Officer Joseph Groves Boxhall
Fifth Officer Harold Godfrey Lowe
Sixth Officer James Paul Moody

Chapter 3

'ALL ABOARD!'

The ship arrived at Southampton during the first hour of 4 April 1912, and by midday men were turning up at the dock to sign on as members of the crew, many of them having transferred to *Titanic* for her maiden voyage because the coal strike had delayed other ships.

On that same day Fifth Officer Lowe sent an unusual postcard entitled 'Irish Greetings' with the phrase printed on in capitals and with spelling mistakes: 'May the Devil never know you'r dead until you'r six months in Heaven.' It was addressed to Nurse Robinson at Moorcliffe Hospital in Sydney, Australia.

Leading Fireman Frederick 'Fred' Barrett (1883–1931) reported:

> Just after departure from Southampton I and about ten other men received orders to empty the coal bunker in boiler room six, where a fire had been discovered.

It took them three days to get it under control, although they could not extinguish it completely, and when Fred went back to see the extent of the damage he noted that the 'bulkhead was damaged from top to bottom, the lower half being warped at the back and the upper half being warped at the front.'

Fireman John Dilley, whose real name was Christopher Arthur Shulver (born 1883), stated:

> From the day we sailed the *Titanic* was on fire, in bunker 6, and my sole duty, together with eleven other men, had been to fight that fire. We had made no headway against it.

Nevertheless, the vessel made a majestic sight as she was moored for five days in berth 43 at dock gate 4, the entrance to the Eastern Dock in

Southampton. The officers greeted people from all walks of life as they began embarking for what was to be the trip of a lifetime. Many were looking forward to starting new lives in the United States. In 1912 class distinction was reluctantly tolerated by most people, and all three classes of accommodation on *Titanic* were segregated by doors and gates, and even without the barriers it seemed to be a known thing that passengers would not enter an area that was not of their class.

One passenger was Edward Pomeroy Colley (1875–1912), a civil engineer and veteran of the Klondike gold rush of the last few years of the nineteenth century in Canada, who was on his way to Vancouver to work as a consultant to the prominent British Columbia industrialist, James Dunsmuir. He came from a well-known Irish family, which descended from kings Edward I and Edward III, right down to the explorer, Sir Ranulph Fiennes (born 1944). His uncle, Major General Sir George Pomeroy Colley (1835–1881) had been killed in action during the British Army's disastrous engagement with Boer commando units at Laing's Nek, in the First Boer (Transvaal) War of 1881. The British were so ashamed that it was the only war for which they did not grant a campaign medal. Nevertheless, he described the ship to his sister-in-law in a somewhat superior manner:

> This is a huge ship. Unless lots of people get on at Cherbourg and Queenstown they'll never half fill it. The dining room is low ceilinged but full of little tables for two, three and more in secluded corners. How I wish someone I liked was on board but then nice people don't sit at tables for two unless they're engaged or married. I wonder my blue blood didn't tell me that? They also have a restaurant where you can pay for meals if you get bored with the ordinary grub. Our most distinguished passengers seem to be W T [William Thomas] Stead, [John Jacob] Astor, Oh, and the Countess of something [Rothes], but her blood is only blue black. (Give me good red corpuscles; I seem to know more about them).

Married couples and families with children in first and second class cabins were allowed to stay together, but in third class all male and female passengers were separated in different cabins. The men and older sons had to sleep in cabins at the bow of the ship, while women, older daughters and young children had to stay in accommodation at the stern.

The lifeboats were situated in sets of four, with four towards the front (bow) of the ship, four towards the back (aft), and four on both the left (port) and right (starboard) sides of the ship. There were sixteen standard wood lifeboats measuring 30 feet long by 9 feet wide, with a capacity of sixty-five persons. There were two wooden cutter 'emergency' lifeboats, which were both just over 7 feet wide, with one being just over 25 feet long and the other 2 feet shorter. Both had a capacity of forty people. There were four Englehardt collapsible lifeboats, which were about 27.5 feet long and 8 feet broad, with a capacity of forty-seven persons. As passengers walked along the promenades they would have noticed the lifeboats, and if anyone knew the details and did the maths they would realise that the total capacity was 1,178 souls, not even half that required in an emergency, which necessitated the total evacuation of the ship. However, as they were spaced out in the way described they may not have noticed the deficiency. The officers certainly did.

According to Fifth Officer Lowe, when speaking at the United States Senate inquiry, it took about ten men to lower each lifeboat from *Titanic*.

> It takes two at each winch. Then there were two jumped in each [life] boat. Then there were some clearing the falls – that is the ropes – and you can roughly estimate it at ten men.

As for the time to lower a lifeboat:

> From the start to finish of putting a [life]boat over, until you get her into the water, it will take you somewhere about twenty minutes.

The officers would welcome on board several multi-millionaires, including American business magnate Colonel John Jacob Astor (1864–1912) and his wife, Madeleine (1893–1940), who boarded at Cherbourg; Benjamin Guggenheim (1865–1912) – who was a member of a very wealthy American family – and his mistress, Léontine Pauline Aubart (1887–1964); American politician Isidor Straus (1845–1912) and his wife, Ida (1849–1912); and another American businessman George Dunton Widener (1861–1912), who travelled with his wife, Eleanor Elkins (1861–1937), and their son, Harry Elkins (1885–1912).

The eight members of the orchestra led by Lancastrian band leader Wallace Hartley (1878–1912) arrived and were given second class cabins. They were required to perform as two separate bands: a quintet and a trio. The quintet performed at teatime and for the occasional after-dinner concerts, while the trio played in the A Le Carte Restaurant and the Cafe Parisian. In all they had to learn the 350 tunes that appeared in the songbook handed out to first class passengers. Wallace wrote in a letter, 'We have a fine band and the boys seem very nice.'

There were at least fourteen men from the world of sport, including tennis players – Karl Howell Behr (1885–1949) and Richard 'Dick' Norris Williams (1891–1968) – and Joseph Bruce Ismay, the chairman and managing director of White Star Line, who had been a talented tennis player in his younger days, having won the Waterloo Championships in Liverpool, the Northern Mixed Doubles Championships in Manchester in 1883, which was later moved to the All England Club in Wimbledon, and the Northumberland Championships in Newcastle in 1884.

About twenty-six couples were on their way to spend their honeymoon in the United States. In second class a businessman named Harry Bartram Faunthorpe (1880–1912) could be seen walking hand-in-hand with a young woman named Lizzie, who he described as his new wife. However, Lizzie – Elizabeth 'Lizzie' Anne Wilkinson (born 1882) was actually someone else's wife.

Also in second class, a few suspicious eyes must also have noticed that Mr and Mrs Marshall did not really look suited to each other. She was a sheepish teenager with a broad Midlands accent and he was a very business-like individual with a southern accent who must have been pushing 40. She was wearing a diamond-encrusted sapphire pendant that caught everyone's eye – just the type of thing an older man might buy for a young mistress perhaps? They said they were on their way to California – hoping that the climate there would help him to get over a recent illness he had suffered; which was exactly the excuse he had given his family – including his wife and daughter. In truth, his name was Henry Samuel Morley (1873–1912), a senior partner in his family's confectionary business in Worcester, and she was Kate Florence Phillips (1893–1964), who had been an employee who worked in one of his sweet shops, and they had begun a secret affair that developed in such a way that Morley was prepared to abandon his family and take Phillips to re-settle on the west coast of America.

Described as a 'handsome bachelor' a Liverpudlian named Joseph Fynney (1876–1912) had often travelled to Montreal, Canada to visit his family, and it is likely that some of the officers who worked on various ships had noticed how strange it was that each time he made the trip he had a young male companion with him. On this occasion it was an 18-year-old lad called Alfred Gaskell (1893–1912), who lived close to the church in Liverpool where Fynney worked with delinquents. They shared a cabin on the same second class ticket, and some passengers must have thought their situation was quite unusual to say the least; there were probably many frowning looks and gossiping tongues.

Talking with a slight French accent, Louis Hoffman led passengers to believe that he was a widower, but officers noticed that he was reluctant to get into conversation with anyone, especially concerning his two boys, Lolo – Michel Marcel Jr (1908–2001) and Momon – Edmond Roger (1910–1953). He was very protective of them and never let them out of his sight. The reason was that his real name was Michel Marcel Navratil Sr (1880–1912) and he was absconding with them from a troubled marriage.

Speaking at the United States Senate inquiry concerning the day of departure, Fifth Officer Lowe stated:

> After the general muster at 8.30 am – on the 10th that was – we manned two boats. Mr Moody, the sixth officer, and myself. On the starboard side, because you must remember that we were laying alongside a wharf, now. And we were sent away in two boats, with two crews, naturally, and we turned around the dock in a row and then came back and got hoisted up, I should say 20 minutes to half an hour. There is not only practice in the rowing of the boats but there is also practice in the lowering away and clearing. We were lowered down in the boats with a boat's crew. The boats were manned, and we rowed around a couple of turns, and then came back and were hoisted up, and had breakfast, and then went about our duties.

The officers were at action stations even before the ship had got out into the open sea, as Edward Pomeroy Colley wrote in a letter:

> We nearly had a collision to start with coming out of Southampton. We passed close to a ship that that was tied up alongside the *Oceanic* and the suction of our ship drew her out of the stream and snapped the ropes that held her and round she swung across our bows! She had no steam up so had to be pulled back by tugs and we had to reverse. The name of her was the New York in case you see it in the papers. It proves conclusively the case of the *Hawke* and *Olympic*.

It was noted at the time that when *Titanic* departed the port Chief Officer Wilde and First Officer Murdoch were at the stern of the ship busy supervising the casting-off of mooring ropes and taking on of tug lines. While *Titanic* was at sea it was among Third Officer Pitman's duties to work out celestial observation and compass deviation, general supervision of the decks, looking to the quartermasters and relieving the officers on the bridge when necessary.

The harbour at Cherbourg was the largest artificial harbour in the world, and *Titanic* sailed into it not long after 6.30 pm on 10 April 1912. Two hundred and eighty-one passengers embarked and an hour and a half later *Titanic* set sail for Queenstown (now Cobh) on the southern coast of Ireland. She dropped anchor at Roches Point outer anchorage in Queenstown and the officers welcomed aboard 123 more passengers, including 14 people from the parish of Aldergoole in County Mayo, who were going to the United States in search of a new life there. As *Titanic* sailed out of the port to begin the journey across the Atlantic Ocean, Fourth Officer Boxhall noticed how strange it was that there were no birds flying around the ship as they usually did when they left a harbour.

Arthur Gee (1865–1912) had intended to sail from Liverpool, but because industrial unrest had affected the port there the ship was delayed and, as fate would have it, he happily agreed to travel to Southampton to board the brand new luxury liner *Titanic*.

He was obviously impressed by the ship, and he told his wife:

> In the language of the poet, 'this is a knock-out.' I have never seen anything so magnificent, even in a first class hotel. I might be living in a palace. It is, indeed, an experience. We seem to be miles above the water, and there are certainly miles of promenade deck. The lobbies are so long that they appear to come to a point in the distance. Just finished dinner. They call us up to dress by bugle. It reminded me of some Russian villages where they call the cattle home from the fields by horn made from the bark of a tree. Such a dinner! My gracious!

However, his dog had acted strangely on the day he made his way to the railway station, as if it sensed foreboding and was trying to warn him not to get on the train. His local newspaper in Lytham in Lancashire reported:

> He kept a dog, which usually reserved its most affectionate demonstrations for Mr Gee's children. Mr Gee, in the course of his business, made frequent journeys from home, but his going and comings were apparently regarded with unconcern by the dog. On the occasion of his departure to embark at Southampton, however, the dog followed the cab to the railway station, and at the station jumped about Mr Gee in so demonstrative a fashion that he remarked on the strangeness of the incident to a friend who was seeing him off, and said how remarkable it was that the dog should appear to know that he was going on a long voyage.

Esther Ada Hart (1863–1928) was travelling with her husband, Benjamin (1864–1912), and their daughter, Eva Miriam (1905–1996). She had never had any kind of premonitions before, but she considered all the talk of the ship being 'unsinkable' was flying in the face of God, and she stated:

> I can honestly say that from the moment the journey to Canada was mentioned till the time we got aboard *Titanic* I never contemplated with any other feelings but those of dread and uneasiness.

She slept during the day and stayed awake at night, presumably thinking that if there was going to be a problem it would happen during the dark hours. Her daughter, Eva, considered that her mother's premonitions were the reason why they survived because they were among the first awake and ready to go to the lifeboats.

On Friday night, 12 April 1912, several passengers on *Titanic* were enjoying a dinner party, as the liner pushed on through the waves. Among them was the newspaper editor William Thomas Stead (1849–1912), who was a great believer in spiritualism. Stories had been in newspapers recently concerning the lid of a sarcophagus in the British Museum, which had once contained the mummy of the sun priestess Amen-Ra. It came to be known as the 'Unlucky Mummy', as misfortune came to almost everyone who came into contact with it, and there were dozens of stories circulating about it. Also at the table was Frederic Kimber Seward (1878–1943), who gave an interview with the *New York World* a few days after the disaster, in which he stated that Stead related stories about the mummy to his fascinated dinner table companions, which he said lasted until after midnight.

It is believed this story began rumours that the mummy of Tcheser-Ka-Ra, the High Priestess Amen-Ra of Thebes was on *Titanic*. Although there was no actual mummy on the ship, and a lid of the sarcophagus, which is said to have once contained the Amen-Ra mummy could be seen in the Egyptian Room at the British Museum prior to *Titanic* sailing, is it possible that the cursed Egyptian artefact was indeed on the ship? It is believed to have been bought by an American, and may have been secreted in one of the general commercial cases such as that of the American Express Company, which included '25 cases of merchandise' – the contents of which were unspecified,

> It was packed carefully, so that no one could guess what its covering-case contained; and arrangements were made that no hitch would be caused by Customs House examinations. And so the coffin was despatched to America on board *Titanic*. It now rests miles deep in the Atlantic. The question which is being discussed is whether the coffin's reputed diabolic power hurled the leviathan to its doom. [See James W. Bancroft, *The* Titanic *Disaster: Omens, Mysteries and Misfortunes of the Doomed Liner*, Pen & Sword Books, 2023.]

Chapter 4

'WE'VE HIT SOMETHING!'

Titanic eventually got out into open water and began her voyage westward, surging relentlessly through the waves and across the North Atlantic trying to make up for lost time, and also the race was on for the westbound Blue Riband accolade, the record highest average speed for a passenger liner crossing the Atlantic Ocean.

Most people were thrilled with anticipation at the prospect of an exciting new future as the rapidly prospering United States of America awaited them. Few passengers harboured foreboding, and it is likely that as they looked around and admired the opulence they would find reassurance in the thought that the designers and builders would not have gone to so much trouble to make it grand and luxurious if they thought for one moment that it could possibly sink.

However, Esther Hart still had no faith in the ship no matter how big and fancy she was. Her forebodings caused her to be particularly wary of how the ship was performing, and she was not convinced when she had been told that *Titanic* was so big that the ship was not supposed to roll badly as she pounded through the waves. She wrote:

> Anyhow it rolls enough for me, I shall never forget it. It is nice weather but awfully windy and cold. They say we may get into New York Tuesday night but we were really due early on Wednesday morning. The sailors say we have had a wonderful passage, up to now there has been no tempest, but God knows what it must be like when there is one. This rough expanse of water, no land in sight and the ship rolling from side to side is very wonderful though they say this ship does not roll on account of its size. We have met some nice people on board and so it has been nice so far. But oh the long, long days and nights it's the longest week I ever spent in my life.

Charlotte Collyer stated,

> I didn't remember very much about the first few days of the voyage. I was a bit seasick and kept to my cabin most of the time. But on Sunday, April 14 [1912] I was up and about. At dinner time I was at my place in the saloon and enjoyed the meal, though I thought it too heavy and rich. No effort had been spared to give even the second cabin passengers on that Sunday the best dinner that money could buy.
>
> After I had eaten, I listened to the orchestra for a little while, and then at nine o'clock or half-past-nine I went to my cabin. I had just climbed into my berth when a stewardess came in. She was a sweet woman who had been very kind to me. I take this opportunity to thank her for I shall never see her again. She went down with the *Titanic*.
>
> 'Do you know where we are?' she said pleasantly, 'we are in what is called the Devil's Hole.'
>
> 'What does that mean?' I asked.
>
> 'That is a dangerous part of the Ocean,' she answered. 'Many accidents have happened near there. They say that icebergs drift down as far as this. It's getting to be very cold on deck so perhaps there is ice around us now.'
>
> She left the cabin and I soon dropped off to sleep. Her talk about icebergs had not frightened me, but it shows that the crew were awake to the danger.

It seems that the stewardess was only trying to spook Charlotte Collyer because the Devil's Hole in the Atlantic Ocean is further south than *Titanic* could have been.

Thomas Andrews was of the opinion that all was shipshape and Bristol fashion:

> Today I made my usual inspection of the ship. I talked to a few people on the deck and then went around the ship to see if there was anyone who could use a little help ... No problems of great importance have risen so far. As passengers are starting to get comfortable in the ship there has been less for me to do. So far, it's been a delightful cruise for everyone.

During the course of the day of Sunday, 14 April 1912, the Marconi wireless operators, John George Phillips, known as 'Jack' (1887–1912) and Harold Sydney Bride (1890–1956) received messages from other ships concerning ice in the area. They had been informed by Captain Smith that icebergs were usually present more to the north of their position, but at 9.00 am in the morning they received a message from Cunard's RMS *Coronia* stating that they had seen '[Ice]Bergs, growlers and field ice.' This was followed up at 11.40 am by a warning from SS *Noordam*, which was also on its way to New York, simply stating that they had seen 'Much ice.' At 1.42 pm RMS *Baltic* warned of 'Icebergs, and large quantities of field ice.' At 7.30 pm Bride overheard a transmission from SS *Californian* to its fellow Leyland Line ship the SS *Antillian* stating that they had seen 'Three large [ice]bergs five miles to the southward of us.'

At 8.00 pm that evening, Fifth Officer Lowe and Third Officer Pitman handed over to First Officer Murdoch and Fourth Officer Boxhall respectively, and they were both due back on duty at midnight. At the same time Sixth Officer Moody began what would be his last watch, and was stationed in the wheelhouse to oversee the helmsman.

At about 9.00 pm Sixth Officer Moody told Second Officer Lightoller that he estimated they should reach the ice region by about 11.00 pm, and both Lightoller and Captain Smith remarked on how quickly the temperature was dropping, yet the weather was clear and the sea unusually calm. It was a starlit night, which Lightoller believed would reflect light off any icebergs that might be nearby. At 9.20 pm Captain Smith, in reference to some iceberg warnings they had received, left the bridge, and gave Lightoller the instructions: 'If in the slightest degree doubtful, let me know.' About ten minutes later Lightoller instructed Moody to telephone the crow's nest and ask the men there to keep a sharp lookout for small ice and to pass the word to subsequent watches.

At 9.40 pm the wireless room received the most serious message to date, which came from the MV *Mesaba* stating:

> Saw much pack ice and great numbers of large icebergs; also field ice. Weather good, clear.

The co-ordinates suggested that the ice field was right in the path of *Titanic*.

At 10.00 pm Second Officer Lightoller handed his duties over to Chief Officer Wilde, and Quartermaster Robert Hichens (1882–1940) – an experienced Cornishman – relieved Quartermaster Alfred John Olliver (1884–1934) at the helm. Therefore, the four officers stationed on the bridge as the ship sailed into the region of the ice flow were the two senior officers, Chief Officer Wilde and First Officer Murdoch, together with Fourth Officer Boxhall and the most junior officer, Sixth Officer Moody.

The final ice-warning message was received at 10.30 pm from *Californian* stating: 'We are stopped and surrounded by ice.' However, although it was obvious that *Titanic* was approaching dangerous waters, this message was not sent to the bridge.

Each watch up in the crow's nest was four hours long; Frederick 'Fred' Fleet (1887–1965) and Reginald 'Reg' Lee (1870–1913) took over the watch from Archie Jewell (1888–1917) at 10.00 pm, who passed on to them the information from the bridge as he had been told to do.

Reginald Lee reported later:

> There was a good deal of haze ahead. It thickened, but the speed of the vessel was not slackened.

Just after seven bells (11.30 pm), Fleet saw a black mass ahead, immediately struck three straight bells, telephoned the bridge, reported to Officer Moody: *'Iceberg right ahead.'* While he was still on the telephone the ship started veering to the port side. However, as they continued to look with bated breath they were alarmed when the crow's nest began to shake as they saw and felt the starboard side of the ship scrape along the jagged edges of the iceberg, and the collision caused ice to fall on the decks. They hoped that the bridge had taken action in time and that it had been a near miss with no severe damage done, so they remained in the crow's nest until being relieved at midnight, 'about 20 minutes later.'

Then Reginald Lee reported what he saw:

> The iceberg appeared to be a dark mass, and the only white spot on it was a fringe along the top. It was first seen at about half-a-mile or less distant.

Able Seaman Joseph 'Joe' George Scarrott (1878–1938) stated:

> It was a beautiful starlight night, no wind, and the sea was as calm as a lake, but the air was very cold. Everybody was in good spirits and everything throughout the ship was going smoothly.
>
> All of a sudden she crashed into an iceberg, which shook the giant liner from stem to stern. The shock of the collision was not as great as one would expect considering the size of the iceberg and the speed the ship was going, which was about 22 knots an hour. I was underneath the forecastle enjoying a smoke at the time. It happened about twenty minutes to twelve o'clock. The shaking of the ship seemed as though the engines had suddenly been reversed to full speed astern. Those of the crew who were asleep in their bunks turned out, and we all rushed on deck to see what the matter was.
>
> We found there was a large quantity of ice and snow on the starboard side of the fore deck. We did not think it very serious so we went below again cursing the iceberg for disturbing us. We had no sooner got below when the boatswain called all hands on deck to uncover and turn all the [life]boats out ready for lowering. We did not think then there was anything serious. The general idea of the crew was that we were going

to get the [life]boats ready in case of emergency, and the sooner we got the job done the quicker we should get below again.

Second Officer Lightoller remembered that about ten minutes after they struck the iceberg,

> ... the Fourth Officer Boxhall, opened my door and, seeing me awake, quietly said, 'We've hit an iceberg.'
> I replied, 'I know you've hit something.' He then said, 'The water is up to F Deck in the mail room.'
> That was quite sufficient. Not another word passed. He went out, closing the door, whilst I slipped into some clothes as quickly as possible, and went out on deck.

Third Officer Pitman had heard a noise and thought the ship was coming to anchor. After a few minutes he decided to go on deck and look around just outside the officers' quarters, where he didn't see or hear anything unusual. He went back inside where he met Second Officer Lightoller, who told him they had 'evidently' hit something. Nevertheless, Third Officer Pitman went back to his bunk, but after a few minutes he got back up, 'as it was no use trying to go to sleep again, as I was due on watch in a few minutes.' While he was putting on his coat, Fourth Officer Boxhall came in and informed him that the mail room was afloat. He asked Boxhall what they had hit and he said an 'iceberg'. After that he put his coat on and went on deck, leaving behind his discharge book and a 'comprehensive collection of stamps'. There Pitman saw men uncovering the lifeboats on the port side of the ship, and he met Sixth Officer Moody on the after part of the deck, and 'I asked him if he had seen the iceberg; he answered, "No, but there was ice on the forward well deck".' Pitman then saw 'a whole bunch' of firemen coming up from the starboard side of the ship.

Fifth Officer Lowe was asleep at the time of the collision, and he was half awakened by hearing voices in the officers' quarters. After a short while he jumped out of bed and looked out of the door, where he saw women in the officers' quarters with lifebelts on. These quarters were usually off-limits to passengers, so, on realising something serious was wrong, he got dressed and went out on deck. There he noticed that the ship was listing slightly at the front.

Identified as M.G.Y., at 11.45 pm *Titanic* sent out a distress call to the SS *Birma*, which was sailing from New York to Rotterdam:

> We have struck iceberg sinking fast come to our assistance – Lat. [latitude] 41.46 N [north] Lon. [longitude] 50.14 W [west].

The co-ordinates where the ship hit the iceberg were actually 41.7325 N (north). 49.9469 W (west).

The starboard side of the mammoth liner had collided with an enormous iceberg, and before the crew could take the necessary evasive action, it had scraped along it and ripped a wide jagged hole which ran for 300 yards along the side, as Thomas Threlfall described: 'The ship had tore herself right open from number 6 section to the fore-hatch', and gallons upon gallons of dark cold water was rushing in. It would seem from the wording of the distress message that only five minutes after the ship hit the iceberg they were aware on the bridge that *Titanic* was in great peril.

Chapter 5

'MAN THE LIFEBOATS!'

Out on deck it was difficult for anyone to be heard because there was a tremendous roar as the ship's exhausts were letting off steam and they had to use hand signals to pass on messages. This carried on for quite some time before it stopped abruptly.

Fourth Officer Boxhall had been somewhere in the vicinity of the officers' quarters when he heard the lookout bell, and as he made his way to the bridge he recalled that he heard First Officer Murdoch give the order 'Hard-a-starboard!' just before the ship's impact with the iceberg. On steamships this meant to turn the wheel to the right to send the vessel in that direction, like a car, although in sailing ships the wheel was turned to the left to make the rudder guide the ship to the right. Then immediately after the collision, the order was given, 'Full speed, astern!' First Officer Murdoch then pulled the lever to close the water-tight doors.

However, the assessment put forward by Professor John Harvard Biles (1854–1933) of the Institute of Naval Architecture soon after the sinking seems to suggest that the weight of the water was not taken into consideration, and if the amount of water that had filled the compartments was enough to cause the ship to list, this could allow water to rush onto the decks, one by one, taking the ship down.

Captain Smith arrived on the bridge, and asked First Officer Murdoch what they had hit, to which Murdoch replied:

> An iceberg, sir. I hard-a-starboarded and reversed the engines, and I was going to hard-a-port round it but she was too close. I could not do any more. I have closed the water-tight doors.

Fourth Officer Boxhall then went along the length of the starboard side of one of the lower decks to try to provide a damage assessment, but he could not see any. However, on C Deck he came upon a man who was carrying a lump of ice and he took it from him. He returned to the bridge and reported that he had not seen any actual damage to the ship. His next mission was to find a carpenter, who told him that the ship was taking on water fast, and a man from the mail room also stated that there was water in the ship's hold. On reaching the mail room, Fourth Officer Boxhall was alarmed to discover that the water: 'was rising rapidly up the ladder and I could hear it rushing in.' He saw mail bags floating around in the water. He reported what he had seen to the men on the bridge, and then he went to the officers' quarters to warn the men who were off duty. Then he heard the disturbing order that they should get the lifeboats ready, so he went up and down the boat deck on both sides of the ship to help with the unlacing and initial preparation of the lifeboats.

When Fifth Officer Lowe reached the boat deck:

> I could feel by my feet there was something wrong – it is not listing it is tipping – she was by the bow; she was very much by the bow. She had a grade downhill ... by the head.

Ida Minahan, who ended up in Lowe's lifeboat 14, stated: 'The frightful slant of the deck toward the bow of the boat [*Titanic*] gave us our first thought of danger.'

Fifth Officer Lowe saw that there was some activity on the starboard side of the ship, so he joined First Officer Murdoch, who was superintending the unlacing and preparation of lifeboats 3, 5, 7 and emergency lifeboat 1. This seems strange as it was the area just above where the ship had hit the iceberg and presumably where water was rushing in. Lifeboat 7 was launched by Murdoch and Lowe at 12.40 pm, and five minutes later they joined Third Officer Pitman and began to uncover lifeboat 5. Some accounts say that Pitman helped with lifeboat 7 but he says Murdoch ordered him to go straight to lifeboat 5, which still had its covers on.

Bruce Ismay 'with his dressing gown and pyjamas on' was also at the scene, and he shouted to Third Officer Pitman,

> Hurry up; there is no time to waste! Fill the [life]boat with women and children!

To which Pitman, not knowing who he was, replied that he would await the captain's orders first. It eventually dawned on him who Ismay was, so Pitman went to the bridge and told Captain Smith what Ismay had suggested, and the captain told him to 'Carry on' – meaning to 'go ahead' with the suggestion.

Many passengers were in two minds whether to get into the lifeboats to begin with. To them the choice was to remain on a massive ship, which was supposed to be 'unsinkable', or dangle in the air in an uncomfortable lifeboat, before being lowered 80 feet down the side of the ship, sail out into the cold dark night, with no heat or light, and even though the sea was unusually calm, it could still get into difficulties.

Able Seaman Scarrott stated,

> Why so many ladies hung back at first was because the experience presenting itself was an awesome one, the mere act of getting into the [life]boats being a difficult one, and the long lowering to the water presenting terrifying prospects.
>
> The first [life]boat to leave the ship was full of firemen, but that was because few ladies were willing to go, and it was imperative to fill the [life]boats. The other members of the crew saved were those required to man the [life]boats, and those who saved themselves at the last moment by jumping overboard to chance being able to float until being picked up. Many more could have been saved if the imminence of the danger had been realised at the time of the first alarm.

On returning to lifeboat 5 Third Officer Pitman and Ismay helped several women and at least two children into the lifeboat. First Officer Murdoch came and told Pitman to get into the lifeboat too and take charge of its evacuation, before asking him to try to stay around the right rear gangway door in case his lifeboat was needed. Murdoch wished him well as he departed.

Third Officer Pitman remembered,

'There was not the slightest suspicion of panic', as he [Murdoch], along with Quartermaster Alfred Olliver, gave the order to lower lifeboat 5. Just before the procedure started, Ismay was becoming over anxious, and on grabbing the falls he shouted to Fifth Officer Lowe: 'Lower away! Lower away!' Lowe did not know who he was and shouted back in a harsh voice, apparently also using expletives,

> If you get the hell out of the way, we'll be able to do something! You want me to lower faster? You'll have me drown the lot of them!

With the first two lifeboats launched – apparently in unsteady stops and starts – First Officer Murdoch and Fifth Lowe uncovered lifeboat 3, and as emergency lifeboat 1 was close by it was dealt with next. It was a smaller 'cutter' lifeboat, which was already uncovered and ready to launch because it was usually used for situations such as to rescue someone who might have fallen into the sea, and it was fitted with a lamp that was lit at 6.00 pm every night. Its capacity was forty, but only twelve people boarded her, the fewest who got into any of the lifeboats that night. There was an identical lifeboat on the other side, emergency lifeboat 2, which was used for the same purpose.

When Fifth Officer Lowe saw that there was nobody else to be seen on the boat deck, he lowered the lifeboat to A Deck, and as he suspended it there, he shouted down to his men to look around to see if anyone was there who could get into the lifeboat, but no passengers appeared so he continued to lower the craft. According to Lowe,

> The davits worked perfectly as expected. Everything went all right, and it could not have been worked better ... With perfect safety.

It was about this time that Fifth Officer Lowe went to his cabin and got his gun, a semi-automatic Browning.

At about one 1.00 pm, Fourth Officer Boxhall reported that he witnessed two mastheads of a steamer in the distance, and Fifth Officer Lowe recalled that somebody mentioned something about a ship on the port bow, and as he glanced over in that direction he saw a steamer with her two mastheads and red sidelights.

Fourth Officer Boxhall went to the chart room and worked out its position, and gave the information to the wireless operators. After reporting to the captain, he sent for some distress flairs and began to fire them into the air, hoping the light of their luminous tail and bursting stars would get the attention of the steamer. He also tried to attract its attention with a Morse (code) light.

Fourth Officer Boxhall informed Chief Officer Wilde that there were no lamps in any of the main lifeboats, so he went to get the lamp trimmer, who in turn, with the help of a few other men, brought as many lamps up on deck as they could find and put them in the lifeboats.

At this time passengers were beginning to appear on the boat deck in substantial numbers. Many of them with anxious faces asked the officers if the situation was serious. The officers could do no more than give then reassuring smiles as they continued with their work. Some passengers wondered why they were sending up distress flares if there was no danger.

Second Officer Lightoller had met Chief Officer Wilde as he came out of the officers' quarters. Wilde told him to accompany him to help him to get the covers off the lifeboats on the port side of the ship, and at about 1.00 am Captain Smith, Chief Officer Wilde and Second Officer Lightoller began to uncover lifeboat 4, which was the first on the port side at the front of the ship to get prepared to launch, and they followed this by uncovering lifeboat 6.

Lightoller stated:

> I commenced stripping off [lifeboat] number 4; then two or three turned up. I told them off to number 4 [life]boat and stood off myself and directed the men as they came up on deck, passing around the boat deck, round the various [life]boats, and seeing that the men were evenly distributed around both the port and starboard.

However, he continued:

> Well, you see, if I may give it to you in the order I was working. I swung out [lifeboat] number 4 with the intention of loading all the [life]boats from A Deck, the next deck below the boat deck. I lowered [lifeboat] number 4 down to A Deck, and gave orders for the women and children to go down to A Deck to be loaded through the windows. My reason for loading through the windows from A Deck was that there was a coaling wire, a very strong wire running along A Deck, and I thought that it would be very useful to tie the [life]boat to in case the ship got a slight list or anything, but as I was going down the ladder after giving the order, someone sung out and said the windows were up. I countermanded the order and told the people to come back on the boat deck and instructed two or three, I think they were stewards, to find the handles and lower the windows. That left number 4 [life]boat hanging at A Deck, so then I went on to number 6.

As Second Officer Lightoller was by lifeboat 6 (at about 1.10 am) he noticed:

> ... she [*Titanic*] was distinctly down by the head, and I think it was while working at that [life]boat it was noticed she had a pretty heavy list to port.

He also noticed that at this time:

> However, having got Captain Smith's sanction, I indicated to the Bosun's Mate [Albert Haines (1880–1933)], and we lowered down the first [life] boat [number 6 at 1.10 am] down to the boat deck, and just at this time, thank heaven, the frightful din of the escaping steam suddenly stopped, and there was a death-like silence.

This caused Chief Officer Wilde to call out the order: 'All passengers over to the starboard side,' which Second Officer Lightoller explained:

> ... was an endeavour to give her a righting movement, and it was then that I noticed the ship had a list. It would have been far more noticeable on the starboard side than on the port.

Captain Smith then gave the order to load women and children into lifeboat 6 and Lightoller began the task by himself. He

> stood with one foot on the seat just inside the gunwale of the boat, and the other foot on the ship's deck, and the women merely held out their wrist, their hand, and I took them by the wrist and hooked their arm underneath my arm.

He appointed Quartermaster Hichens to take charge of the vessel.

When the 67-year-old American millionaire, Isidor Straus, was offered a place in lifeboat 8 he refused saying, 'I do not wish any distinction in my favour which is not granted to others.' When his wife of forty years, Ida, was asked to join a group of people who were waiting to get into the lifeboat she said,

> I will not be separated from my husband. As we have lived, so we will die – together.

Then she gave her fur coat to her English maid, Ellen Bird (1881–1949), and insisted that she get into a lifeboat. As the lifeboats were being rowed away from the stricken vessel the couple were seen to be consoling each other and awaiting their inevitable fate. Straus's English manservant, John Farthing, was also a victim. It was filled to less than half of its capacity.

From about 1.20 am Chief Officer Wilde, Second Officer Lightoller, Fifth Officer Lowe and Sixth Officer Moody began working on lifeboats 12, 14 and 16 at the rear of the ship on the port side. It must have crossed their minds by this time that some areas in the depth of the ship must have been completely filled with water and the dreadful thought hit them that without doubt there had to be some instances of people having drowned.

Second Officer Lightoller recorded:

> Between one [life]boat being lowered away and the next [life]boat being prepared, I usually nipped along to have a look down the very long emergency staircase leading direct from the boat deck down to C Deck. Actually built as a short cut for the crew, it served my purpose now to gauge the speed with which the water was rising, and how high it had

got. By now the fore deck was below the surface. That cold, green water crawling its ghostly way up that staircase was a sight that stamped itself indelibly on my memory. Step by step it made its way up, covering the electric lights, one after the other, which, for a time, shone under the surface with a terrible weird effect.

The men stood back to allow the women to pass, except in one or two cases when men tried to rush, but they were very soon stopped. This occurred at the [life]boat I was in charge of, number 14. About half-a-dozen foreigners tried to jump in before I had my complement of women and children. They could not understand the orders I gave them, but I drove them back with the [life]boat's tiller. One man jumped in twice and I had to throw him out the third time.

Shortly afterwards the fifth officer, Mr Lowe, came and took charge of the [life]boat. I told him what had happened. He drew his revolver and fired two shots between the [life]boat and the ship's side into the water as a warning to any further attempts of that sort. When our [life]boat was lowered we had fifty-four women, four children (one of them a baby in arms), one sailor, two firemen, three stewards, and one officer; total sixty-six souls. There was a man in the [life]boat who we thought was a sailor but he was not. He was a window cleaner (William Harder [1872–1947]).

Able Seaman Scarrott recalled:

Directly I got in my [life]boat [number 14] I jumped in, saw the plug in, and saw my dropping ladder was ready to be worked at a moment's notice; and then Mr Wilde, the chief officer, came along and said, 'All right, take the women and children,' and we started taking the women and children. There would be 20 women got in the [life]boat, I should say, when some men tried to rush the [life]boats, foreigners they were, because they could not understand the order which I gave them, and I had to use a bit of persuasion. It seemed no sooner the order came along my side of the deck from Mr Wilde – I heard him personally give the order – we had just started to get people into the lifeboat. We heard that before we got abreast of our [life]boat, we heard it further along the deck, and I continued getting the women in, and when Mr Wilde came along he gave the order again and assisted me to get the women into the [life]boat.

Charlotte Collyer stated:

Mr Lowe was very young and boyish-looking, but somehow he compelled people to obey him. He rushed among the passengers and ordered the women into the [life]boat. Many of them followed him, in a dazed kind of way.

Noticing that five lifeboats had left the ship without having an officer in them, Fifth Officer Lowe said to Sixth Officer Moody: 'An officer ought to go in one of these [life]boats', and asked him who he thought it should be. Probably taking into consideration the fact that Lowe had made himself the most familiar with the lifeboats during the practices than any of the other officers, Moody is said to have told Lowe to get in the lifeboat, so Lowe then boarded lifeboat 14 and took command.

And so, under Fifth Officer Lowe's orders, lifeboat 14 was lowered away from *Titanic*'s boat deck at around 1.25 am. Fearful of the consequences that might happen if people tried to jump into the lifeboat as it passed the lower decks on its descent to the sea, Lowe fired his gun three times alongside the side of the ship to warn people to stay back. As it neared the water a very dangerous situation occurred when, for some reason, the back end of the lifeboat stopped lowering while the bow continued until it reached the water – causing the stern of the lifeboat to dangle about 5 feet in the air. With the stern stuck in the air, the crew were forced to release the ropes attached to the lifeboat to lower it from the ship – causing the lifeboat's stern to come crashing down into the sea. The crash seems to have caused a leak that allowed about 8 inches of water to fill up the bottom of the lifeboat.

From about 1.30 am First Officer Murdoch and Sixth Officer Moody went to the rear starboard of the ship to work on lifeboats 9, 11, 13 and 15. By this time the passengers were realising the severity of the situation and these lifeboats were filled to capacity or even more. A surviving passenger, Hilda Mary Slayter, who was in lifeboat 13, launched about an hour after the ship began to sink, remembered:

> Of all the heroes who went to their death when the *Titanic* dived to its ocean grave, none deserved greater credit than the members of the vessel's orchestra. The orchestra played until the last. When the vessel took its final plunge the strains of a lively air, mingled gruesomely with the cries of those who realized that they were face to face with death. From the moment the vessel struck, or as soon as the members of the orchestra could be collected, there was a steady round of lively airs. It did much to keep up the spirits of everyone and probably served as much as the efforts of the officers trying to prevent panic.

At this time Second Officer Lightoller was busy getting the stranded lifeboat 4 to a position where it could be filled with passengers. Madeleine Astor, wife of American millionaire, John Jacob Astor, got into the lifeboat, and when her husband asked if he could join her because she was in a delicate condition, Second Officer Lightoller refused. Then the Ryerson family –Emily Maria Borie (1863–1939), her daughters Susan 'Suzette' Parker (1890–1921) and Emily Borie Jr, and her 13-year-old son, John 'Jack' Borie (1898–1986) – got in. Her husband, Arthur Larned Ryeson (1851–1912), stayed on the deck. However, when Lightoller saw the boy he tried to remove him, but Arthur Ryerson pointed out that he was just a boy and Lightoller relented.

Emergency lifeboat 2 was not far from lifeboat 4, and Second Officer Lightoller saw a group of men trying to take over the vessel. He regained control by threatening them with an empty gun. At 1.45 am Chief Officer Wilde asked Fourth Officer Boxhall to stop sending the flares and to accompany him to the front port side and they began to get emergency lifeboat 2 ready for launching. Fourth Officer Boxhall stated,

> Every time I fired a signal I had to clear everybody away from the vicinity of the socket, and then I remember the last one or two distress signals I sent off [life]boat 4 had gone, and they were then working on the collapsible [life]boat which was on the deck.

He remembered that he was sending rockets up right until the time he was sent away in the emergency lifeboat.

Chief Officer Wilde told Fourth Officer Boxhall to get into the craft and take charge, and it was lowered down towards the water a few minutes before 2.00 am. First Officer Murdoch and Chief Officer Wilde then went to the aft port side of the ship and began to take charge of lifeboat 10 – the last regular lifeboat to be launched.

The officers were beginning to become wary of the fact that some passengers were starting to panic, and Second Officer Lightoller stated:

> It was about this time that the chief officer [Wilde] came over from the starboard side and asked, did I know where the firearms were ... I told the chief officer, 'Yes, I know where they are. Come along and I'll get them for you', and into the first officer's cabin we went – the chief, Murdoch, the captain [Smith] and myself – where I hauled them out, still in all their pristine newness and grease. I was going out when the chief shoved one of the revolvers into my hand, with a handful of ammunition, and said, 'Here you are, you may need it.'

At 2.00 am it was time to deal with the four collapsible lifeboats, the last possibility of salvation, and Chief Officer Wilde and First Officer Murdoch went to the rear starboard side to prepare collapsible lifeboat C, while Second Officer Lightoller went to the port side to get collapsible lifeboat D ready.

Dinghy D was lifted and hooked to the tackles where emergency lifeboat 2 had been. The crew then formed a ring around the craft and allowed only women to pass through. The lifeboat could hold forty-seven, but only fifteen women stepped on board, so Second Officer Lightoller allowed men to take the vacant seats, and Michel Navratil handed his two boys to someone in the lifeboat. Then Colonel Archibald Gracie (1858–1912 – died in 1912 but not in the *Titanic* disaster) arrived with more female passengers and all the men immediately stepped out and made way for them; a gallant move which sealed their fate. Chief Officer Wilde asked Lightoller to go with the lifeboat, but he retorted: 'Not damn likely' and he stepped back on deck. While the collapsible lifeboat was lowered to the ocean, two men were seen to jump into it from the rapidly flooding A Deck.

Harold Bride stated:

> Then came the captain's [Smith's] voice: 'Men, you have done your full duty. You can do no more. Abandon your cabin now. It's every man for himself. You look out for yourselves. I release you.'
> That's the way of it at this kind of time, every man for himself. I looked out. The water was then coming into our cabin while he worked. I looked out. The boat deck was awash. Phillips clung on sending and sending. He clung on for about ten minutes, or maybe fifteen minutes after the captain had released him.

Harold Cottam (1891–1984), the wireless operator on *Carpathia*, received the last message sent by Jack Phillips from *Titanic*, which was 'Come quickly; engine room filling to boilers.'

The situation was becoming desperate, and they still had collapsible lifeboats A and B to get off, as the water rose as high as the boat deck. Second Officer Lightoller climbed on to the top of the officers' quarters, stripped the covers and cut away the ropes with a penknife. He was able to send collapsible lifeboat B down to the flooded deck, but it flipped over and Harold Bride became trapped under it. As Lightoller moved to the opposite side of the ship to deal with collapsible lifeboat A, *Titanic* took a great plunge forward, and several people realised that it had started to go down very quickly, so Lightoller turned to face the sea and dived in. He had started to swim clear when he was sucked against the grating of one of the large ventilator shafts, and he was taken down with the ship as she slipped under the surface. As the water hit the still hot boilers, the blast blew him back to the surface where he found himself alongside the capsized collapsible lifeboat B. He and Bride grabbed on to it. As *Titanic* went down, the forward funnel broke loose and toppled his way, narrowly missing him.

However, this turned out to be his salvation, and saved him from being sucked down with the ship, as he stated in his *I Was There* interview:

> A bit later the forward funnel guise carried away, and the funnel, weighing perhaps 50 or 60 tons, fell down with a crash on the water and just missed the raft with some of us hanging onto it, by inches. There were a good many it didn't miss. The wash of the falling funnel had evidently picked us up, raft and all, and flung us clear of the ship altogether.

As he struggled desperately he wondered what was making it so difficult for him to keep his head above the water, until he realised:

> ... it was my great Webley revolver, still in my pocket, that was dragging me down. I soon sent that on its downward journey.

Knowing the ship was sinking fast, Chief Officer Wilde, First Officer Murdoch and Sixth Officer Moody were all showing devotion to duty to the last as they tried to get the last collapsible dinghy released and launched. Suddenly, the ship took a great surge forward and the front disappeared beneath the surface of the waves. As she sank further and further it either sucked them down, or left them with no chance of surviving in the icy water. If any of their bodies were recovered they were not identified.

Dr Henry Washington Dodge stated:

> The officers in charge of loading the boats were cool and masterful, preventing as far as possible all disorder and enforcing the command to care for women and children first.

This was a reflection on what became known as the Birkenhead Drill.

An at-a-glance list of the sequence of events concerning the officers who took some part in the launching of the lifeboats:

Time	Place	Lifeboat (Number)	Launchers (Officers)
12.40 am	Starboard side for	Lifeboat 7	Murdoch, Lowe
12.45 am	Starboard side for	Lifeboat 5	Murdoch, Lowe,

Time	Side	Lifeboat	Officers
12.55 am	Starboard side for	Lifeboat 3	Pitman * Murdoch, Lowe
12.55 am	Port side for	Lifeboat 4	Wilde, Lightoller (unsuccessfully)
1.00 am	Port side for	Lifeboat 8	Wilde, Lightoller
1.05 am	Starboard side for	Lifeboat E1	Murdoch, Lowe
1.10 am	Port side for	Lifeboat 6	Lightoller
1.20 am	Port side aft	Lifeboat 16	Wilde, Lightoller, Moody
1.25 am	Port side aft	Lifeboat 14	Wilde, Lightoller, Moody, Lowe *
1.30 am	Port side aft	Lifeboat 12	Wilde, Lightoller
1.30 am	Starboard side aft	Lifeboat 9	Murdoch, Moody
1.35 am	Starboard side aft	Lifeboat 11	Murdoch, Moody
1.40 am	Starboard side aft	Lifeboat 13	Murdoch, Moody
1.40 am	Starboard side aft	Lifeboat 15	Murdoch, Moody
1.45 am	Port side for	Lifeboat E2	Wilde, Boxhall *
1.50 am	Port side for	Lifeboat 4	Lightoller
1.50 am	Port side aft	Lifeboat 10	Murdoch, Wilde
2.00 am	Starboard side	Lifeboat CC	Murdoch, Wilde
2.05 am	Port side	Lifeboat CD	Wilde, Lightoller
2.10 am	Starboard side	Lifeboat CA	Murdoch, Wilde, Moody
2.10 am	Port side	Lifeboat CB	Lightoller *

Port side – left side
Starboard side – right side
For – forward
Aft – aft
* – officer left the ship
E – emergency
C – collapsible

Chapter 6

PERIL ON THE SEA

Fourth Officer Boxhall and the other rowers got emergency lifeboat 2 about 100 yards from the stricken ship and then allowed the craft to drift for a while. Then he heard someone call out from *Titanic* through a megaphone, asking them to bring emergency lifeboat 2 back towards the starboard stern of the ship. When they got to about 200 feet from the stern, Boxhall could detect that there was some suction from *Titanic* so he instructed the rowers to pull the lifeboat away from the ship in a north-easterly direction.

When asked if he heard any cries from the stricken passengers, Boxhall replied:

> Yes, I heard cries. I did not know when the lights went out that the ship had sunk. I saw the lights go out, but I did not know whether she had sunk or not, and then I heard the cries. I was showing green lights in the [life]boat then, to try and get the other [life]boats together, trying to keep us all together.

Third Officer Pitman kept lifeboat 5 near the gangway as he had been asked, but when it did not open they rowed away for about 100 yards to avoid being caught in any suction that might be created as the ship went down. Lifeboat 7 was still close by with Lookout in Charge George Alfred Hogg (1883–1946) and Lookout Archie Jewell, and as the two lifeboats got close together they decided to transfer four passengers from lifeboat 5 to lifeboat 7.

Third Officer Pitman saw lifeboat 3 being lowered, and he watched on in horror as the front of the massive ship got lower and lower in the water and the different rows of lights disappeared under the surface, but he said that he did not give up hope until he saw the last line of lights on the forecastle head disappear. When asked if he knew the time when *Titanic* disappeared from the surface he replied:

> Two-twenty, exactly, ship's time. I took my watch out at the time she disappeared, and I said, 'It is 2.20 [am]', and the passengers around me heard it.

It must have been heartbreaking to hear the cries coming from the drowning people, and although the officer suggested that they go back several passengers were against it and they persuaded him not to.

He remembered that at about this time he saw what he thought was a white light from the stern of a sailing ship about 5 miles away. He stated that the passengers and crew in his lifeboat behaved well, but they did not return to rescue drowning passengers for fear of being swamped. When they got about 200 yards from the starboard side of where *Titanic* sank they stopped rowing and rested on their oars.

The second in command of lifeboat 14, Able Seaman Scarrott recorded:

> The aft fall got twisted and we dropped the [life]boat by releasing gear and got clear of the ship. When the [life]boat was in the water we rowed clear of the ship. There were four men rowing. We then saw four other [life]boats well clear and fairly well-filled with women and children. We went to them and found none of them had an officer in charge. So the fifth officer [Lowe] took charge of the lot, ordering them to keep with him. The *Titanic* was then about fifty yards off, and we lay there with the [life]boats. Mr Lowe was at the helm.
>
> The ship sank shortly afterwards, I should say about 2.20 am on the 15th, which would be about two hours and forty minutes after she struck. The sight of that grand ship going down will never be forgotten. She slowly went down bow first with a slight list to starboard until the water reached the bridge then she went quicker. When the third funnel had nearly disappeared I heard four explosions, which I took to be the bursting of the boilers. The ship was right up on end then. Suddenly she broke in two between the third and fourth funnel[s]. The after part of the ship came down on the water in its normal position and seemed as if it was going to remain afloat, but it only remained a minute or two and then sank. The lights were burning right up until she broke in two. The cries from the poor souls struggling in the water seemed terrible in the stillness of the night. It seemed to go through you like a knife.
>
> Our officer (Lowe) then ordered all the [life]boats under his charge to row towards where the ship went down to see if we could pick up anybody. Some of our [life]boats picked up a few. I cannot say how many. After that we tied all our [life]boats together so as to form a large object on the water which would be seen quicker than a single [life]boat

by a passing vessel. We divided the passengers of our [life]boat amongst the other four, and then taking one man from each [life]boat so as to make a crew we rode away amongst the wreckage as we heard cries for help coming from that direction.

Lifeboat 14 had met up with lifeboats 4, 10, 12 and collapsible lifeboat D. Fifth Officer Lowe decided that his lifeboat should return to the site of the wreck to try to rescue survivors. He began to transfer all the passengers out of lifeboat 14 into the other lifeboats in the group. As the woman and children were being transferred, Lowe was unimpressed to find that there was a man in the lifeboat disguised as a woman, so he 'pitched' him into the other lifeboat. Two crewmen, Able Seaman Frank Oliver Evans (1884–1952) and Able Seaman Edward John Buley (1885–1917), were transferred from lifeboat 10 into lifeboat 14.

Thomas Threlfall was in lifeboat 14, and recalled:

> Then he (Officer Lowe) called to several other [life]boats close by, 'Throw out your painters,' and we linked them all up. Mr Lowe passed about fifty women and children from his [life]boat, and said, 'We will go for the wreckage', to which other people were clinging. From the wreckage we picked up four men. Then Mr Lowe called out, 'There's a [life]boat over there and she's sinking.' Although we were then towing a collapsible [life]boat with about eighty people in her we reached the sinking [life]boat just as the water was up to her gunwale and took twenty-six men and one woman, a Mrs [Rhoda Mary] Abbott [1873–1946], off her. I held the woman in my arms till we reached *Carpathia*.

Second class passenger Clear Annie Cameron (1877–1962) noted:

> Officer Lowe decided to go back to search for survivors, and transferred all the passengers from lifeboat 14 into lifeboat 10. According to Nellie [her friend Ellen 'Nellie' Wallcroft (1875–1949)] a 'madman' kept shaking the [life]boat and they feared it would capsize, so he was pushed overboard. Lifeboat 14 was reportedly the only one which went back to look for survivors and six victims were pulled from the water, but two of them died.

After Fifth Officer Lowe allowed a first class passenger named Charles Williams into the lifeboat to help to row, some people expressed a fear that they might be swamped by desperate people if they returned, so he reluctantly agreed to wait until the screams from those in peril had died down a bit before he went back to help the desperate survivors. Lifeboat 14 made the journey of approximately 150 yards back to the disaster site, and when they arrived at the wreckage they were confronted with the dreadful sight of countless dead bodies floating in the icy water.

Able Seaman Scarrott remembered:

> When we got to it the sight we saw was awful. We were amongst hundreds of dead bodies floating in lifebelts. We could only see four alive. The first one we picked up was a male passenger. He died shortly after we got him in the [life]boat. After a hard struggle we managed to get the other three. One of these we saw kneeling as if in prayer upon what appeared to be a part of a staircase. He was only about twenty yards away from us but it took us half-an-hour to push our [life]boat through the wreckage and bodies to get to him; even then we could not get very close, so we put out an oar for him to get hold of and so pulled him to the [life]boat. All the bodies we saw seemed as if they had perished with the cold as their limbs were all cramped up. As we left that awful scene we gave way to tears. It was enough to break the stoutest heart.

Ida 'Daisy' Minahan (1879–1919), an American passenger, did not speak favourably of Fifth Officer Lowe. She testified that at first he did not want to go back for any survivors, and was persuaded to do so by some of the women in the lifeboat. She also alleged that he swore at her and said she suspected that he had been drinking. However, her testimony is not supported by any of the other passengers, and it seems that she was mistaken about his decision to go back. It is understandable that the officer would have had little patience with any of the passengers who were being awkward or inconsiderate – and there were some – and he may well have sworn in the heat of the moment.

Ellen Wallcroft stated:

> Officer Lowe wanted to go back to the rescue, but the women begged him not to go.

Lily May Futrelle (1876–1967), a surviving passenger, was of the opinion that Fifth Officer Lowe was,

> A leader with a cool head, desperate courage, and a knowledge of the sea, who rescued people with his own hands.

During the Board of Trade inquiry, Sir Robert Finlay summed up with the words:

> [Fifth Officer Lowe] waited, an operation of the most painful character, requiring great nerve and great coolness, he waited until the sea had done its work with the great majority of people, and then put back and picked up a few of the survivors. But to suggest, as some questions which were put suggested, that there was inhumanity in not pushing into the crowd of drowning people in the hope of saving them, is, I submit, a course based upon ignorance of the fundamental conditions that attend on the endeavour to save people who are struggling in the water when, if you push your [life]boat among them, the only result will be that those in the [life]boat are added to the roll of victims.

The final person to be rescued was a steward named Harold Charles William Phillimore (1888–1967), who was clinging to the top of some wreckage that looked like a piece of staircase. They had to push their way through many dead bodies to get to him. After rescuing all the survivors they could find in the water, Fifth Officer Lowe constructed a makeshift sail which allowed them to make more progress. As they did so, they saw collapsible lifeboat D and sailed towards it. A rope was thrown to the craft and it was taken in tow.

Second Officer Lightoller took charge of the thirty men on the upturned collapsible lifeboat B, and Harold Bride estimated it would take *Carpathia* 'an hour or so' to get to them. Regular lifeboats 4, 10 and 12, eventually began to separate from Lowe's craft, and at about 4.30 am as daylight began to appear, Lightoller saw lifeboats 4 and 12 drifting around in the distance, so he took out a whistle and used it to gain the attention of the people in the other two lifeboats, calling for them to row over to him. He and some of the men from the upturned lifeboat got into lifeboat 12 and the rest got into lifeboat 4, while collapsible lifeboat B was left drifting in the water. First class passenger Algernon Henry Barkworth (1864–1945) was transferred from the collapsible craft to lifeboat 12, and he remembered:

> With daylight, a strong breeze arose which threatened to submerge us. When we were rescued the water was up to our knees.

The cable ship *Carpathia* was on its way from New York to Gibraltar and fortunately was in the region, and on receiving a distress signal from *Titanic* it immediately set a heroic course towards the disaster area. After working through dangerous ice fields it arrived at the scene at 4.00 am on the morning of 15 April 1912. All the lifeboats spent the rest of the night drifting in the sea about 1 mile from the disaster site, until *Carpathia* came into sight at dawn.

Fourth Officer Boxhall spotted *Carpathia* on the horizon just as it started to get light at 4.00 am, and guided her to the lifeboats with a green flare.

Able Seaman Scarrott expressed his relief when he realised that they were going to be rescued:

> Just then we sighted the lights of a steamer, which proved to be the *steamship Carpathia* of the Cunard Line. What a relief that was. We then made sail and went back to our other [life]boats. By this time day was just beginning to dawn.
>
> All our [life]boats proceeded towards *Carpathia*. She had stopped right over where our ship had gone down. She had got our wireless message for assistance. When we got alongside we were got aboard as soon as possible. We found some survivors had already been picked up. Everything was in readiness for us – dry clothes, blankets, beds, hot coffee, spirits, etc, everything to comfort us.
>
> I must say that the passengers when they were in the [life]boats, especially the women, were brave and assisted the handling of the [life]boats a great deal. Thank God the weather was fine or I do not think there would have been one soul left to tell the tale. The last of the survivors were got aboard about 8.30 am. The dead bodies that were in some of the [life]boats were taken aboard and after identification were given a proper burial.
>
> We steamed about in the vicinity for a few hours in the hope of finding some more survivors, but we did not find any. During that time wives were enquiring for husbands, sisters for brothers, and children for their parents, but many a sad face told the result.
>
> *Carpathia* was bound from New York to Gibraltar, but the captain decided to return to New York with us. We arrived there about nine pm on Thursday the 18th. We had good weather during the trip, but it was a sad journey. A list of the survivors was taken as soon as we had left the scene of the disaster.
>
> On arrival at New York everything possible was ready for our immediate assistance – clothing, money, medical aid and good accommodation. In fact, I think it would have been impossible for the people of America to have treated us better.

Carpathia arrived back in New York on the evening of 18 April 1912, where the passengers were taken off at Pier 54, but the crew were taken to Pier 61 by a tender named *George E. Starr*. There seems to have been an attempt to keep the crew away from awkward questions until they had been briefed.

Third Officer Pitman sent a message home which simply announced: 'Safe – Bert.' Fourth Officer Boxhall complained that he was in pain, and three days later he was examined by a doctor in Washington DC.

Chapter 7

'*TITANIC* DISASTER: GREAT LOSS OF LIFE'

First reports of the disaster were very sketchy, and confused many of the families who were waiting for information concerning their loved ones. However, after about a week, more accurate and informative narratives began to appear in newspapers around the world, such as the following report from *The Melbourne Argus* for 22 April 1912:

> *Titanic* Disaster – Death Roll Number 1,635 – Survivors' Thrilling Narratives – Men Shot While Rushing Boats – Explosion Breaks Ship in Two – Average Speed of 21 Knots:

A story of awful loss of life and terrible suffering is told by the survivors of the *Titanic* disaster. According to the official figures, issued by the White Star Line Shipping Company, 1,635 people perished, and 795 were saved. More than 100 of the survivors are receiving treatment in hospital.

The *Titanic* plunged before sinking, and as the vessel dived the decks assumed such a steep incline that hundreds were thrown downwards. Eight hundred people leaped into the sea as the vessel went down. An explanation has been given of the vessel breaking in two. The ice and water entered through a hole made by the collision, and, reaching the boilers, caused a terrific explosion. There were regrettable instances of men – mostly foreigners – attempting to rush the [life]boats. As a result, several were shot dead by the officers.

The story that Captain Smith committed suicide is incorrect. He behaved with great valour, and just before the vessel sank shouted to the passengers who had not been rescued, 'Be British!' In one [life] boat the Countess of Rothes [1878–1956], and several other women took the places of unskilled stewards at the oars. As a result of the disaster, Atlantic liners have been ordered to take a course 100 miles south of that followed by the *Titanic*.

Phillips, the wireless operator, stuck to his post. He continued sending calls for help until the dynamo ceased. 'Come quickly! Engine room filling to boilers,' was the last wireless message from the *Titanic* received by the *Carpathia*.

An investigation of the disaster by the United States Senate Committee has been begun, in giving evidence, Mr Ismay, the managing director of the White Star Company, stated that the vessel was not pushed to its speed limit. The average speed was 21 knots an hour. Mr Ismay has bitterly protested against the attitude of the Senate Committee.

Among the saved are seven infants whose names are unknown. As more details are obtained from survivors regarding the loss of the White Star liner *Titanic*, the picture of the disaster becomes more awful. The official figures show that more than 1,000 people perished, while of the 795 who were saved a large number are in hospitals. From the stories of the saved an idea can be formed of the terrible sufferings of the women – many of them very thinly clad – who spent hours in the open [life]boats in bitterly cold weather. While the behaviour of the crew was excellent, there were instances of passengers who tried to force their way into the [life]boats before all the women had been provided for. This led to the use of revolvers, and a number of deaths resulted. Among the most affecting scenes at the landing of the rescued passengers from the *Carpathia* on Thursday night was the sight of the women steerage passengers. Thinly clad and shivering with the cold, their eyes red with incessant weeping over the loss of husbands, families, and friends, they formed a pathetic group as they came mournfully ashore. Preparations had been made by different charitable organizations to provide for their immediate wants, and they all met a sympathising welcome, and were speedily succoured.

Official figures have been issued by the White Star Line company regarding the fate of the passengers and crew. The figures are as follow: Dead ... 1,635; Saved ... 795.

Captain E.J. Smith, who was on the bridge, ordered that all on board should take life savers [life jackets]. The [life]boats were lowered at once The first contained more men than women, as the men were the first to reach the deck. When the women appeared, the rule that women and children must be saved first was strictly observed. The officers drew their revolvers, to be ready to maintain order on board, but they were not used in most cases.

The *'New York World'* publishers detailed stories of the disaster told by some of the surviving passengers. They say that the iceberg, which was 80 feet high, was sighted when the *Titanic* was a quarter of a mile away. When the steamer crashed into the [ice]berg the engines were at once stopped and the bulkhead doors closed almost simultaneously by levers operated from the bridge.

Mrs [Leila] Meyer [1886–1957], wife of Edgar J. [Joseph] Meyer [1884–1912], of New York, whose husband went down with the steamer, pleaded with her husband to be allowed to remain with him, but he refused and threw her into the lifeboat, reminding her of their nine-year-old child at home.

In the course of his narrative Mr [Lawrence] Beesley [1877–1967], the science teacher at Dulwich College, states that he saw a [life]boat half-full of women on the port side. A sailor asked Mr Beesley if there were any more ladies on his deck, and there being none, invited Mr Beesley to jump in.

There were no officers in the [life]boat. No one seemed to know what to do. The [life]boat was swung under another descending [life]boat, but with great promptitude a stoker in Mr Beesley's [life]boat cut her away, thus preventing the other boat from falling on and crushing her. The stoker then took charge of the [life]boat.

Large numbers of the rescued were absolutely destitute. Philanthropic persons have provided 210 of the saved with outfits. American newspapers, in commenting on the disaster, are unanimous in paying tributes to the valour and discipline of the crew. They complain that there were not sufficient [life]boats, and also protest against the vessel being driven at the rate of 21 knots in the region of the icebergs.

Of the survivors 140 are suffering so acutely from the effects of the exposure and injuries sustained that they are receiving treatment in the hospital. An explosion occurred shortly before the vessel went down. When the rafts were full, several people begged piteously for help, but the crew, for the sake of self-preservation, had to refuse to permit more to come aboard. Most of the persons on the rafts prayed throughout the weary night.

The *Titanic* plunged before sinking, and many of those on board jumped overboard. Some were picked out of the water by persons in the [life]boats and saved. A number of the passengers also embarked in collapsible [life]boats, and were subsequently picked up. When dawn broke it was seen that the sea was strewn with the dead. There are rumours of atrocities by the frenzied members of the crew. Among other stories, it is alleged that passengers in the [life]boats were shot, and that several people who were swimming to the [life]boats were brained. The majority of the witnesses, however, do not confirm these accounts.

There is no doubt, in the light of stories told by the survivors, that a large number of people lost their lives owing to their implicit confidence in the *Titanic* remaining afloat. Several passengers retired to their beds, so convinced were they that the vessel would not sink. Others stood at the rail, ridiculing as 'land-lubbers' those who were taking to the [life]boats.

Agonising horrors were presented just before the steamer sank. The last scenes occupied only a few minutes. After the ship split the stern rose until the decks were like a steep incline. Scores leaped overboard, and hundreds of others scrambled upwards, madly clutching at anything to prevent themselves being precipitated into the sea.

As the incline became steeper hundreds lost their foothold, and tobogganed downwards, shrieking piteously as they fell. Hundreds who escaped being drawn down in the vortex clung to wreckage and rafts. It is believed that 900 jumped from the ship as it was sinking. Only the hardiest could withstand the water, owing to its being icy cold. In most instances even the good swimmers, after a few vigorous strokes, were chilled by the intense cold, and their stiffened forms floated away. The [life]boat which was commanded by the purser capsized, owing to three women rushing to the side, to say farewell to relatives. There were 30 women on this [life]boat, and it is believed that none of them survived.

The immediate cause of the steamer going down was the explosion of the boilers. The impact with the iceberg made a hole in the starboard side, through which ice and water were admitted. This water reached the boilers, which exploded, and the explosion was so great that the steamer was broken in two. Some of those who were saved died from exposure while being transferred to the *Carpathia*.

When a steamer sinks there is frequently an explosion at the last moment, followed by a great rush of steam and smoke, which would at first sight suggest that, with the flooding of the stokeholds, the boilers had exploded. This theory, however, is not universally accepted by engineers.

What is more probable is that pent up bodies of air in the hull gradually reached enormous pressure with the inrush of the water, and that this rends the decks, and with the consequent weakening of the whole structure, the hull itself may break in two. A rush of cold water in the heater furnace plates would set up severe strains through unequal expansion and contraction, but with new boilers and the high factor of safety that is always provided on the great liners, it is regarded as unlikely that the boilers would explode from this cause.

It is reported that a mad rush for the [life]boats was made by a number of foreigners. When this was threatened the first mate, Murdoch stood with his revolver in his hand and shouted 'I'll kill the first man who rushes the [life]boat.' Three of the foreigners took the risk. They ran forward, but as they did so Murdoch fired. Two of them dropped dead, one being shot through the head and the other's jaw being shot away. The third man was felled by the quartermaster.

Lady [Lucy Christiana] Duff-Gordon [1863–1935], wife of Sir Cosmo [Edmund] Duff-Gordon [1862–1931], who was among those saved, has made a statement that she saw a man shot, and his body fell into the [life]boat in which she was. Several others who attempted to rush the [life]boats were felled. Many of the passengers, she said, died quickly from exposure to the icy cold. The body of the man who was shot remained in the [life]boat until the survivors were picked up by the *Carpathia*. In order to secure a place in the lifeboats six Chinese [passengers] hid themselves under the seats [of the lifeboat] prior to the launching. They were not detected. President [William Howard] Taft [1857–1930] has been informed that Major [Archibald] Butt [1865–1912], of the White House military staff, shot 12 persons during the struggle which took place, and was then himself shot.

A graphic description of the collision and the foundering is given by three French survivors. They state that they were playing cards when they heard a violent noise, resembling the sound made by racing screws. They saw the ice rubbing the vessel's side. The ship took a tremendous list, and there was a momentary panic, but it quickly subsided and confidence was restored. Addressing the passengers who had hurried on deck, the captain [Smith] said, 'Let everyone put on a lifebelt. It will be more prudent.' Shortly after the collision the steamer's band started to play popular airs to reassure the passengers. It is asserted by the Frenchmen that at first none of the passengers wanted to leave with the [life]boats, as they believed that there would be no risk in remaining aboard the *Titanic*. Hence some of the [life]boats left with comparatively few passengers.

'Our [life]boat,' the survivors added, 'was rowed a distance of about half a mile from the steamer. The spectacle was fairylike, the *Titanic* stationary and illuminated, resembling a fantastic stage picture.' Suddenly the lights were extinguished and there was an immense clamour. The air resounded with supreme cries for help, and shrieks of anguish. The *Titanic* sank quietly. The suction was imperceptible from our [life]boat but there was a great backwash.

'After our [life]boat left we saw a group of passengers launching, with difficulty, a collapsible [life]boat. About 30 clambered into it, and the result was that the light craft gradually filled with water.' The majority of these passengers were either drowned or they perished from exposure. 'The crew of the *Titanic* acted with sublime self-sacrifice. Much useless loss of life would have been avoided but for the blind faith of a number of passengers in the vessel's unsinkableness.'

The account of how the wireless operator, Phillips, stuck to his post is a story of great heroism. He remained at his instrument, sending away the calls for help, until the dynamo ceased to work. His assistant, Bride, was washed overboard, and subsequently saved. Both of Bride's feet were injured. He bears testimony to the courage displayed by the chief wireless operator, Phillips, and states that he did not cease signalling until the water came in and stopped the dynamo. While Phillips was informing the *Olympic* that the steamer was sinking Bride strapped a life belt on him. The water then entered the wireless room. A stoker attempted to remove the life belt from Phillips, but the assistant struck him down. Phillips then went aft. The assistant found a collapsible [life]boat lying on the deck, and clung to it as it was washed overboard. The sea was then dotted with people who were depending upon life belts. A great stream of sparks was rushing from the steamer's funnels.

Eventually Bride was picked up by one of the [life]boats. Another wireless states that he saw Phillips make his way to a raft, where he died from exposure. The operator on the *Carpathia* who received the wireless messages from the *Titanic* is named Cottam. He states that Phillips's last message was – 'Come quickly; engine room filling to boilers.'

An explanation has been given of the neglect on *Carpathia* to reply to enquiries made by the warships of the United States navy and by the president (Mr Taft). It is stated that answers were not sent because the operator of the wireless was suffering from physical exhaustion.

The story that Captain Smith and the chief engineer had committed suicide after the vessel struck has been denied. It has now been ascertained that the graphic description of how Captain Smith, after a revolver had been taken from him in the library had shot himself through the head, also the account of the suicide of the chief engineer, were the imaginings of a passenger who had lost his reason. Nearly all the rescued commend the captain for his heroism. They state that he was literally washed from his post. When the [life]boats were clear, Captain Smith, addressing the crew, said – 'Men, you have done your duty. You can do no more. It's every man for himself.' Just, before the *Titanic* went down, the captain, using the megaphone, shouted two words 'Be British' to the mass of men on the deck. Later the captain [Smith] was seen helping those who were struggling in the water. He refused to take advantage of an opportunity to save himself. Mrs George Widener [Eleanor Elkins Widener], wife of the millionaire, is one of the rescued. Her husband was drowned. She states that she saw Captain Smith jump from the bridge into the sea. A moment before she saw another officer commit suicide by blowing out his brains with a revolver.

Colonel Archibald Gracie, in describing what occurred after the collision and before the *Titanic* foundered, said: 'I was driven to the topmost deck and saw no other survivor. After an immense wave had swept the liner, I grasped the brass hand railing desperately but was forced to release my hold when the ship plunged. I went down and was whirled around for what seemed an indeterminable time. I was then taken under the water. When I eventually came struggling to the surface, I seized a wooden grating, which was floating nearby, and hung on till I had recovered breath. Then I discovered a large canvas cork raft near to me, and, with another survivor, I struggled to it and climbed on it. We then both did what we could to rescue others who were floundering in the water. When dawn broke there were thirty of us on the raft standing knee deep in the water, and still afraid to move, lest the raft and we should be all cast again into the sea. So we stayed four long and terrible hours before we were picked up by the boats from the *Carpathia*.'

One of the passengers gives a thrilling account of the manner in which the fifth officer (Mr Lowe) assisted in the work of saving many lives, and warning people not to jump and thus swamp the [life]boats. He says that when Mr Lowe's collapsible [life]boat was launched he hoisted the mast and sail, and then collected all the other collapsible [life]boats afloat, arranged an adequate crew for each. Then he connected them by lines, so that they were all moving together. Having accomplished this, he returned to the wreck, and saved another collapsible [life]boat containing 30 persons, all scantily clad. This craft was on the point of sinking.

The Countess of Rothes, who is among the saved, is an expert oarswoman. She practically commanded her [life]boat. When it was found that the men could not row properly several of the women took the places of the weak, unskilled stewards.

Action has already been taken with a view to keeping other vessels clear of the ice regions. The American hydrographic office has ordered the lanes for Atlantic liners to be moved 180 miles south of the track taken by the *Titanic*. All steamers which have already started on the westward journey from England to North America since the *Titanic* disaster have been ordered to travel further south.

It has been reported in the cable messages that Mr George Eastman [1854–1932], president of the Eastman Kodak Company, is amongst the missing passengers on the *Titanic*. The Melbourne office of Kodak (Australasia) Limited on Saturday received a cable message from the head office of Mr Eastman's company, stating: 'Eastman here. No Kodak people on *Titanic*.'

An investigation has been begun by the Merchant Marine Committee of the Senate into the reason for the inability of the officers of the *Titanic* to provide [life]boat accommodation for all on board. It has already been announced that every survivor who could throw light on the question had been subpoenaed to attend the inquiry. Joseph Bruce Ismay, managing director of the White Star Line (the owners of the *Titanic*), was the first witness called. He stated that he was asleep when the vessel collided with the iceberg. The *Titanic* was not pushed to its speed limit, but it averaged about 21 knots an hour. Only once did he consult the captain regarding the vessel's movements, and it was then arranged that they should not attempt to reach New York before 5 o'clock on Wednesday morning. He entered the [life]boat only when there was no response to the call for women. He left about an hour after the collision occurred. So far as he could judge, the iceberg struck the steamer between the bow and the bridge.

The second officer of the *Titanic*, Lightoller, maintained that the [life]boats were well filled, preference being given to women. He accounted for the saving of so many of the crew of the fact that of six people picked out of the water five were firemen or stewards. There was no demonstration aboard the steamer, not even lamentations among those who were left behind. He dived as the vessel sank and was sucked under. While holding fast, close to a ventilating blower, a terrific gust through the blower, probably caused by a boiler explosion, blew him clear of the sinking steamer. He reached the surface near a [life]boat, and was rescued.

The committee re-examined Mr Ismay and the officers of the *Titanic*, who gave evidence that the steamer followed strictly the southernmost track for west-bound vessels. She had encountered no ice previously, and was proceeding with vigilant lookouts at full speed, but with reduced consumption of fuel. The speed was probably 21 or 22 knots.

Lightoller, the second officer, stated that when the first [life]boat was lowered the deck was 70 feet above water, but when the last put off it was only a few feet above the waterline.

The committee at first refused to allow Mr Ismay or any of the officers or crew to return to England, but afterwards decided that [only] Mr Ismay, four officers and twelve of the crew should remain.

Mr Ismay, when interviewed, said that the committee had been brutally unfair. 'My conscience is clear' he said. 'I took the chance to escape when it came. I did not seek it.'

Mrs [Madeleine] Astor (whose husband, Colonel John Jacob Astor, the well-known millionaire, was one of those lost), in speaking of the wreck, says that she hazily remembers when amid the confusion she was about to be placed in the [life]boat that her husband stood by her side. She has no knowledge of how he died.

Varying accounts have been given as to what happened to Colonel Astor, the millionaire, and to Mr W.T. Stead, the editor of the '*Review of Reviews*', before they were drowned. Several survivors think they both reached a raft, but that they succumbed from the cold and dropped off.

Mr Charles Melville Hays [1856–1912] (president of the Canadian Grand Trunk Railway system), was not saved, as has been reported, but went down with the steamer. When last seen he was standing on the *Titanic* bidding farewell to those who were going off in the [life]boats.

Mr Joseph Bruce Ismay, when rescued, was clad in pyjamas, slippers and an overcoat. He was dazed by the cold. Several narratives show that he rendered active assistance in filling and lowering the [life]boats. Among the saved was Charles Williams, the racquets champion.

Colonel Archibald Gracie states that when he reached the surface he found the second officer [Lightoller], and the son [John 'Jack' Borland Thayer III (1894–1945)] of Mr J.F. [*sic*] Thayer [John Borland Thayer II (1862–1912)], the railway magnet, swimming alongside him. One of the funnels fell, scattering the bodies which were in the water.

Among the lost are Mr [Henry Forbes] Julian [1861–1912], the well-known metallurgical engineer, Mr [Howard Brown] Case [1863–1912], managing director of the Vacuum Oil Company, and the Reverend Ernest Courtenay Carter [1858–1912], vicar of St Bude's, Whitechapel, and his wife [Lillian Carter (1867–1912)].

Among the rescued are seven infants, whose names are unknown. One was suffering from scarlet fever, and another from meningitis. Several of the survivors are seriously ill with pneumonia, which they contracted owing to the bitterly cold weather. An infant son [Hudson Trevor Allison (1911–1929)] of Mr [Hudson Joshua Creighton] Allison [1881–1912], the Montreal banker, was the sole survivor of a family of four.

Mrs [Mary Graham Carmichael] Marvin [1894–1975], who was on her honeymoon, was prostrated when, on coming ashore, learned that her husband [Daniel Warner Marvin (1894–1912] was among those who had drowned. He placed her in the lifeboat which carried her to safety, and exclaimed, 'It's alright, little girl; you go. I will stay.' When the [life] boat in which she was pushed off from the steamer he threw her a kiss; that was the last time she saw him.

Mr Ismay has given instructions that the liners of the International Mercantile Marine are to be equipped with lifeboats and rafts sufficient for all aboard. The House of Representatives has adopted a measure to provide for the shelling of icebergs by warships. It is claimed that this will afford target practice as well as clear dangers from the path of navigators. Widespread sympathy is expressed for the sufferers by the disaster, and the relatives of those who lost their lives. Congress has adjourned as a token of respect.

The tragedy furnished a curious instance of a life saved by superstition. Mrs Fison, a Vancouver lady, who was visiting her friends in England, having read in some almanac purporting to forecast the year's events that a great sea disaster was likely to occur in the middle of April, refused to return to Canada during that month, though urged by her husband to take her passage from Southampton by the *Titanic*.

The last scenes, as depicted by some of the survivors, presented an agonising array of horrors. As the leviathan split at the stern she rose precipitately, and scores, believing that the end had come, leaped overboard as though to anticipate their fate, or in the desperate hope of finding some wreckage to cling to. Others scrambled madly upwards towards the stern; but as the incline steepened hundreds lost their footing and tobogganed downward to destruction, shrieking pitiably. Hundreds who escaped the vortex clung to wreckage and rafts, but were quickly killed by the cruel exposure. It is believed that eight hundred jumped from the vessel as she sank. Only the hardiest, however, could withstand the icy water. A few vigorous strokes and their stiffened forms floated away.

A [life]boat commanded by the purser capsized. Through the folly of three female occupants rushing to the side of the boat to bid farewell to relatives left behind on the *Titanic*, they caused the craft to overturn. It is believed that not one of the thirty women included among its occupants survived.

Chapter 8

CONCISE BIOGRAPHICAL TRIBUTES

Henry 'Harry' Tingle Wilde was born on 21 September 1872 in West Derby, Liverpool, and grew up in Walton, Liverpool. He was the younger of two sons in the family of four children born between 1860 and 1872, of Henry Wilde (1838–1872) of Ecclesfield, Sheffield, and his wife, Elizabeth (née Tingle, 1835–1881) of Loxley, which was in the Bradfield district of Sheffield. They had married in 1860. Henry Wilde was employed as an insurance surveyor. Henry Tingle Wilde was christened at the Loxley Congregational Chapel on 24 October 1872, and became known as 'Harry'. Henry Wilde died of hepatitis on 6 June 1872, just a few months before he was born, and his mother died when he was aged 9. He was 6 feet 1 inch tall, with a dark complexion, blue eyes and dark-brown hair.

He served his four year apprenticeship with the James Chambers Company of Liverpool aboard SS *Greystoke Castle*, from 23 October 1889 to 22 October 1893; but he remained on board as third mate until 5 December 1893. Ten days later he took the examination for second mate, and after passing for the qualification he received the certificate on 8 January 1894. His address at the time was given as 155 Rice Street, Walton, Liverpool.

He joined the Liverpool and Maranham Steamship Company as a third officer on 15 February 1894, and served on the cargo ship SS *Hornsby Castle*, which was re-named SS *Brunswick* in 1895, Wilde being brought aboard her on 23 March of that year. He seemed to fare well on steamships and he was appointed second mate on 17 June 1895, and first mate on the following 29 July. He served on *Brunswick* until April 1896. On 30 June 1896 he signed-on to the SS *Europa* as second mate, for three voyages which lasted about three months each.

55

He joined the White Star Line on 16 July 1897, his first ship being the SS *Cevic*, and in February 1898 he gained a qualification to 'Render first aid to the injured' with the St John's Ambulance Association.

He married a Liverpool girl named Mary Catherine 'Polly' Jones (1863–1910) on 3 August 1898, in the Welsh Calvinistic Methodist Chapel on Princes Road in Liverpool. They had Jane Elizabeth 'Jennie' in 1900, Henry 'Harry' Owen in 1904; George Arnold (known as Arnold) in 1906, Annie in 1909, and Archie and Richard in 1910. They were all been born in Liverpool.

He joined SS *Tauric* in April 1898, as fourth officer, and then the SS *Delphic* in September 1898, which served the New Zealand trade, and between October 1899 and May 1900 he served as second officer aboard the *Cevic* again. He passed his extra master's on 9 July 1900. His next ship was the SS *Persic* as second officer for the Liverpool to Sydney run. However, he was not taking to the long voyages to Australia and he told his sister-in-law he did not 'want to come out this way anymore,' and he was working on *Cedric* at the time of her maiden voyage on 11 February 1903, under the command of Captain Herbert Haddock. During his final month aboard *Cedric* in November 1903, his wife would have been alarmed to read in the newspapers that there had been a collision between his ship and SS *Titian*. However, the reports proved to be wrong.

He entered the Royal Naval Reserve on 26 February 1902, as a sub-lieutenant with the service number 001471, and he was promoted to lieutenant on 24 June 1905. Wilde served on a number of White Star Line ships, mainly on the Liverpool to New York, and Australia routes, including the *Arabic* in 1905, RMS *Celtic* from 1905 until 1906 and SS *Medic* from 1906 to 1908.

A terrible tragedy hit the family in 1910, while they were living at 25 Grey Road, Walton. Wilde's wife, Polly, was pregnant with twins, who were to be named Archie and Richard, when she contracted scarlet fever. The children were born on 10 November 1910, but Archie died aged 14 days and Richard died aged 25 days. Polly Wilde succumbed to the disease on Christmas Eve that year. They were buried together at Kirkdale Cemetery, Liverpool. In January 1911 Wilde altered his will to make sure his four young children were cared for in the event of his death, and he appointed members of his family as executors and guardians.

He served as chief officer of *Olympic* from 8 July 1911 to 1 April 1912, while he was still living at 25 Grey Road (which still exists). He was chief officer of *Olympic* on 20 September 1911 when she collided with *Hawke*. The ship was involved in two more incidents when it struck a sunken wreck and had to have its damaged propeller replaced in February 1912, and it nearly ran aground while leaving Belfast. However, he remained on *Olympic* until his last voyage on her to New York from 13 to 30 March 1912.

In a letter written by Charles Lightoller in 1935, he referred to Henry Tingle Wilde, who was a fellow Lancastrian,

> He was a fine fellow and one for whom I had the greatest admiration. It was frightfully hard luck on him that he should have been temporarily transferred from *Olympic* to *Titanic* for just one voyage.

He is remembered on the family grave at Kirkdale Cemetery in Liverpool, which is marked by an obelisk and gravestone. The full inscription reads in capital letters:

> In loving memory of Mary Catherine (Polly), the dearly beloved wife of Lieutenant Henry T. Wilde, RNR, who departed this life 24 December 1910, aged 36 years. Also the twin sons of the above, Archie and Richard, who died in infancy December 1910. A loving mother and a faithful friend. Also Captain [sic] Henry T. Wilde, RNR, Acting Chief Officer, who met his death in the SS [sic] Titanic Disaster, 15th April 1912, aged 38 [sic] years. One of Britain's Heroes.

***

William 'Bill' McMaster Murdoch was born of a well-known seafaring family on 28 February 1873 at 3 Sunnyside in the parish of Urr, Dalbeattie near Dumfries – in what was then Kirkudbrightshire – Scotland. He was the second of three sons, and fourth of seven children to Captain Samuel Murdoch (1843–1917), a master mariner, and his wife, Jane 'Jeannie' – née Muirhead (1839–1914). They were married at St Michael's Church in Liverpool, and lived in Oldham Street, Liverpool, before moving to Dalbeattie. His older brother, James, had been born in England, and his sister, Mary, died aged 4 in 1869. His parents and other members of the family are buried in Dalbeattie Cemetery. He was educated at Dalbeattie Primary School, and then at the high school, and at the time of the 1881 census the family home was briefly at 341 High Street, Urr, Dalbeattie, until they moved to Oakland Cottage in the High Street, which Samuel Murdoch had built himself. William Murdoch is remembered as being 'an intelligent and hard-working scholar', being particularly 'very good in mathematics', and he graduated with top honours in 1887. As a boy he enjoyed the hobby of constructing model ships.

He was apprenticed to William Joyce and Company of Liverpool for five years, serving his apprenticeship aboard the barque *Charles Cotesworth*, with which he passed his second mate's certificate at his first attempt in 1893. He was assigned to the barque *Iquique* and served under his father. When in Liverpool he lodged at 86 Upper Stanhope Street in the city.

The extra master's certificate was the highest qualification for a nautical officer at that time, which Murdoch gained at his first attempt, at Liverpool in 1896, being the only one of his *Titanic* officers to pass all the Board of Trade examinations at the first attempt. He was recorded as being 5 feet nine inches tall, with a fair complexion, brown hair and hazel eyes.

He joined William Joyce and Company in 1897, serving aboard the barque *Lydiate*, sailing to America and China, before he signed off in 1899. During the Boer War he trained as a lieutenant in the Royal Naval Reserve. He entered the Royal Naval Reserve in 1902, with the service number 001401, and from 1901 to 1907 six members of his family were lost at sea.

He met a New Zealand school teacher named Ada Florence Banks (1873–1941) on his way back to England, and they married by special license at St Denys Church in Southampton, on 2 September 1907. They lived at the home of Captain William James Hannah and his wife; the captain having come from a family of seafarers who were natives of Kirkcudbrightshire, and he was assistant marine superintendent of White Star Line. They eventually settled at 94 Belmont Road, Portswood, Southampton.

He joined the White Star Line in 1900, his first job being to sail with Charles Lightoller on the Liverpool-Cape Town-Sydney route aboard *Medic*, one of the five Jubilee Class of ocean liners.

Murdoch gradually rose in rank while serving on several White Star Line ships. These included serving as second officer on *Arabic* in 1903, for the Liverpool-Queenstown-New York City route. During a trip across the North Atlantic one dark night, another ship was seen bearing down on *Arabic*. His superior, Officer Fox, ordered the ship to steer hard-to-port, but realising this was the wrong order Murdoch ran into the wheelhouse, and brushing the quartermaster aside, he took over the wheel and kept the ship on a straight course. His professional judgement averted a collision by a matter of inches. Among other ships he also served from 1907 to 1911 on RMS *Adriatic*, one of the 'Big Four' with *Celtic*, *Cedric* and *Baltic*; sometimes serving with Joseph Boxhall; and in 1911 he joined *Olympic* with Captain Smith and several crew members who would later transfer to *Titanic*.

There has been rumours that William Murdoch was also married to a Liverpool girl named Elizabeth Crogent (1868–1939), and had seven children to her. However, research revealed that the man she married was not the *Titanic* officer. A check of the Lancashire Online Parish Clerk records of the Banns of Marriage at St Bride's Church in Toxteth,

Liverpool – available on the internet – reveals that Elizabeth Crogent actually married William Edward Murdock (born 1867) on 2 September 1892. The age, middle name, surname and address do not match.

On 20 April 1912, the second day of the US Senate inquiry, while wireless operator Harold Bride was giving evidence, a very upset young woman came into the room and demanded to know information about First Officer Murdoch. No one had the nerve to tell her his fate, and eventually Senator William Alden Smith called to Charles Lightoller, 'Would you be good enough to tell this lady whatever she wishes to know.' Second Officer Lightoller took her to one side, and shortly afterwards she left.

The true identity of the 'weeping woman' was not ascertained, but on 21 April 1914 the *New York Tribune* reported,

> The appearance of Bride, who was brought in an invalid chair, has feet swathed in bandages, his ankles having been crushed, added a dramatic feature to the session, as did the appearance of a young woman, said to be a Miss Harding, who sobbingly inquired for Second Officer Lightoller, from whom she sought some further tidings of the First Officer Murdoch, who went down with the ship.

On 4 July 1907, while he was serving aboard *Adriatic,* Murdoch wrote a letter from Queenstown to a 'Miss Nancy' and it has been suggested that the recipient of the letter was the woman who went into the courtroom. The letter has survived.

The four surviving officers sent a letter, written by Charles Lightoller, to Ada Florence Murdoch from the Hotel Continental in Washington DC on 24 April 1912:

> Dear Mrs Murdoch,
>
> I am writing on behalf of the surviving officers to express our deep sympathy in this, your awful loss. Words cannot convey our feelings – much less a letter.
>
> I deeply regret that I missed communicating with you by last mail to refute the reports that were spread in the newspapers. I was practically the last man, and certainly the last officer, to see Mr Murdoch. He was then endeavouring to launch the starboard forward collapsible [life] boat. I had already got mine from off the top of the quarters. You will understand when I say that I was working the port side of the ship, and Mr Murdoch was principally engaged on the starboard side of the ship, filling and launching the [life]boats.
>
> Having got my [life]boat down off the top of the house, and there being no time to open it, I left it and ran across to the starboard side,

still on top of the quarters. I was then practically looking down on your husband and his men. He was working hard, personally assisting, and overhauling the forward [life]boat's fall. At this moment the shop dived, and we were all in the water. Other reports as to the ending are absolutely false. Mr Murdoch died like a man, doing his duty. Call on us without hesitation for anything we can do for you.

Yours very sincerely,

C.H. Lightoller, 2nd Officer
J. Groves Boxhall, 4th Officer
H.J. Pitman, 3rd Officer
H.G. Lowe, 5th Officer

While researching his book, Walter Lord formed the impression that First Officer Murdoch and Second Officer Lightoller had great respect for each other, and during the 1936 BBC radio broadcast entitled I *Was There* the listener can detect a warmth of feeling he had for Murdoch, who he described as an 'old shipmate' in his memoirs.

First Officer Murdoch was played by Scottish actor Ewan Stewart in the 1997 James Cameron film *Titanic*, in which Murdoch was incorrectly portrayed as having committed suicide by shooting himself. The people of Dalbeattie requested an apology from the makers of the film. On 15 April 1998, Scott Neeson, the executive vice president of Twentieth Century Fox, arrived in Dalbeattie, where he donated funds for a number of prizes in William Murdoch's name. However, no 'full apology' was ever given.

As a point of interest, a similar situation happened concerning the 1964 motion picture *Zulu!*, when Private Alfred Henry 'Harry' Hook VC (1850–1905) was portrayed as an insubordinate soldier who drank from a bottle of spirits. His descendants objected to this as he was known to be a good soldier – and a teetotaller. He was awarded the Victoria Cross for valour in the battle.

William Murdoch is named on the family gravestone at Dalbeattie Cemetery, he is commemorated on a stone plaque in the wall of Dalbeattie Town Hall, and the Murdoch Memorial Prize was established at his old school and is competed for annually. Articles retrieved from the *Titanic* wreck area in 2000 have been linked to First Officer Murdoch.

***

Charles 'Charlie' Herbert Lightoller was born in Chorley, Lancashire on 30 March 1874, the youngest of seven children to Frederick 'Fred' James Lightoller (1842–1913), and his wife, Sarah Jane – née Widdows (1843–1874), who had married in Blackpool in 1863. He was baptised on 24 April

1874. Having already lost a child named James at birth, his mother died of scarlet fever soon after Charles Lightoller was born, followed by two of his siblings, Richard and Caroline. He had three surviving sisters, Jane, Gertrude and Ethel.

The Lightoller family was involved in the cotton spinning industry and the Lightoller Mill on Standish Street was well-known in the town. Frederick Lightoller gave up his life as a captain in the army to take over his father's cotton mills. They lived at 'The Firs' in Yarrow, Chorley, which was one of the first houses to have electric lighting in the town.

Lightoller's father married Margaret Barton in 1876. Their marriage was childless, and Margaret died in 1881. Frederick Lightoller had an affair with the family maid, Joyce Gladwin, resulting in the birth of a daughter named Janet in 1883. Frederick Lightoller together with Joyce, Janet and Jane moved to New Zealand, leaving his other children, Charles, Gertrude and Ethel, with family in England, although he remained in contact with them.

Charles Lightoller attended Chorley Grammar School. He was said to be a bright boy with great potential, but he was apparently difficult to manage, and when he was 13 he asked if he could go to sea. His life at sea was one of the most eventful on record.

He became apprentice to the William Price Line of Liverpool in 1888, and was soon aboard a large ship named the barque *Primrose Hill*, bound for San Francisco. His next voyage was on the barque *Holt Hill*, which lost its mast in a storm in the South Atlantic, and was forced to put into Rio de Janeiro during a revolution and a smallpox epidemic. After makeshift repairs, she lost her new mast in another storm in the Indian Ocean, and in 1889 ran aground on St Paul, an uninhabited island in the Indian Ocean. The chief mate was killed in the shipwreck, and after eight days the survivors were rescued by the *Coorong* and taken to Adelaide in Australia, where they spent Christmas 1889.

An interesting report about this incident appeared in the Perth *Daily News* on 7 April 1933, under the title '*Titanic* Survivor's Adventure (From Our Correspondent) in London.'

> A romantic quest for treasure believed to be hidden on the bleak volcanic island of St Paul, in the Indian Ocean, is being planned by a man who was shipwrecked at that desolate spot more than 40 years ago. His wife and two daughters will accompany him. The quest is of interest to Australia, because the principal adventurer, when rescued from the island, was taken in an Australian ship to Adelaide.
>
> The man is Commander Charles Lightoller, hero of half a dozen shipwrecks, and the only officer saved of those remaining in the *Titanic*

when she went down with the loss of hundreds of lives. Many sailors have tried to find the treasure that pirates hid on St Paul after they had raided the old East India merchant boats centuries ago. Commander Lightoller believes he knows where the treasure is likely to be. Telling of his plans, Commander Lightoller said he was shipwrecked and nearly died of starvation on St Paul. 'It was in 1889, when I was an apprentice,' he said. 'I was serving in a four-masted barque, the skipper of which was a Liverpool man who boasted that he never let a ship pass him. We were bound for the East Indies, and were running light when a big sailing ship hove in sight and was gradually overtaking us. Our skipper gave the order for every sail to be set and soon we ran the other boat out of sight.

'But the fact that we had nearly been overtaken so upset the skipper that even when we had left the other ship behind he would not give orders to take in any sail. We knew we were close to land, and at eight o'clock that night, when the second mate came on duty, he had some of the sail taken in. A gale was blowing, and we could not see where we were heading. All of a sudden the rain lifted, and we saw that we were tearing straight for some cliffs at 14 knots. It was too late to do anything and we crashed, with the bow held firmly between two pinnacles of rock. Luckily only one man was killed, and the 42 who were left on board scrambled down a rope on to the island. There was no chance to rescue anything, and the ship was soon battered to pieces. We found that we had only one match between us, so we started a fire, and never allowed it to go out.

'The eight days we spent there were a nightmare. The volcano was still in action, and there was hardly anything to eat – just a few penguins and rabbits. To get water we had to climb 2,000 feet and then carry it back in our boots or oilskin pants.

'We found 42 uninhabited huts and a number of boats, so old that when we touched them our fingers went through the wood. They were what the pirates had used when they raided the merchant ships on the East Indies route. I saw a beautiful lagoon and a number of wrecks near it, but we could not get anywhere near. It is in a cave near the lagoon that I am certain there is rich treasure to be found. I was getting on the track of it when we were rescued.

'On the eighth day we were ravenous, and so weak that we could hardly crawl about. We knew food was to be found on the island, but it was not until 12 years later that I discovered where it was hidden. It was in the afternoon of the eighth day that we sighted the *Coorong*. She had a crew of only six, and they themselves were running short of food. After some hesitation they took the 42 of us on board, and we set sail for Adelaide, 22 days away. Most of that time we lived on sugar and water, but we thought ourselves lucky to get that.'

Lightoller signed on with the clipper ship *Duke of Abercorn* for his return to England. His third voyage was again on *Primrose Hill*, this time to Calcutta in India. On this voyage they survived a cyclone. In Calcutta Lightoller passed his second mate's certificate. While serving as third mate on the windjammer *Knight of St Michael* the cargo of coal caught fire. For his successful efforts in fighting the fire and saving the ship, Lightoller was promoted to second mate.

He obtained his mates ticket. He left the windjammers and joined Elder Dempster's African Royal Mail Service starting a career on steamships. After three years on the West African coast, he nearly died from a serious bout of malaria.

Lightoller went to the Yukon in Canada in 1898 to prospect for gold in the Klondike gold rush, and he had a brief stint as a cowboy in Alberta. He worked his passage back as a cattle wrangler on a cattle boat, and arrived back in England penniless in 1899. He obtained his master's certificate and joined Greenshields and Cowie, where he made another trip on a cattle boat, this time as third mate of the *Knight Companion*.

He joined the White Star Line in 1900, and his first assignment was as fourth officer on *Medic*, a passenger cargo liner on the Britain-South Africa-Australia run. William Murdoch served with him, but after one voyage, he was switched to the Atlantic routes, mostly on *Majestic*, his first employment under Captain Smith.

While travelling to Australia on *Medic* he met Iowa Sylvania Zillah Hawley-Wilson, a 17-year-old Australian teenager known as 'Sylvia', who was returning home to Hell's Hole in Sydney after a year of education in England. They became a couple and were married at St James's Church in Sydney on 15 December 1903. Her father was a gold miner, and her mother was the first female sanitary inspector of Sydney, who assisted in stamping out the terrible plague of 1900 and she was one of the founders of the Hospital Saturday Fund in Australia.

In 1907, Lightoller was promoted to third officer on *Oceanic*, and in the same year the home port of *Oceanic* was changed from Liverpool to Southampton, which meant another move for the Lightoller family. From second officer on *Oceanic*, Lightoller moved up to first officer on *Majestic*, and then moved back to *Oceanic* as its first officer. At the time of the *Titanic* disaster they were living in a residence at Hound Walk in Netley Abbey near Southampton.

He was called to testify at the American and British inquiries into the *Titanic* disaster, and he was stringently questioned in both cases. As the most senior surviving officer he found himself having to defend Captain Smith, the other officers and White Star Line against some of the more serious charges brought against them.

The Sydney Daily Telegraph on 18 June 1912 published the following, under the title 'The Lost *Titanic* – Senior Surviving Officer's Story – Letter from Mr H Lightoller':

'The whole horror can only be just lived down, and it will take time,' says Mr Herbert Lightoller, the senior surviving officer of the ill-fated *Titanic*, in an interesting letter to his mother-in-law, Mrs C Hawley-Wilson, of Pitt and Barlow Streets in Sydney.

'I wonder if you know that I had sailed as second of the *Titanic*. On the last afternoon before we sailed Mr Wilde came as chief, instead of taking command of the *Cymric*, which was laid up owing to the coal strike. So for the voyage Murdoch went back first, I went second, and Blair (the proper second) stayed ashore.

'You would be worrying a good deal if they sent the rank of survivors without the names. Several of my relatives in New York, were misled owing to the statement that the first officer [Murdoch] was lost. I'm not going into any details of the affair. For one thing they do no good, and for another it is all too recent. I am content to a certain extent in the knowledge that I did my duty. And all I could, under the circumstances. I also have the satisfaction of having gone down with the ship and taken my chance with the rest who were compelled to do the same. That I survived, I suppose, will be attributed to my unfailing good luck.

'To me it seems a year since I left home, instead of three weeks. There being no one saved from the purser's staff, and not one engineer, also being the senior surviving officer has made it pretty warm. I finally had to engage a stenographer both in Washington [DC] and New York, to cope with the correspondence. Then I also had solo charge of all the men, who managed to almost create an international affair through appealing to the British Embassy. This matter also I managed to clear up; so, on the whole, I have had my time occupied.

'The inquiry in Washington [DC] was a farce – so far as obtaining facts to assist the laws of the country. Of what use is it to the Legislature to know whether the cries were groans or loud or weak cries for help, etc, and all the other morbid details, which merely pandered to the sensational press? Of course, when we get home it will all have to be lived again, notwithstanding that we have given it all to the United States Senate, and gone through it again before the British Consul and Crown Lawyer.

'The accident itself was caused by a combination of circumstances which would not occur again, perhaps, in a hundred years. A perfectly calm night, an absolutely smooth sea, no moon (between last and first quarter), the blue side of the berg towards us. Had there been the slightest breeze, the ripples breaking on the berg would have given a phosphorescent glow; a swell would have done likewise; a moon would have given reflected light; the white side would have discovered it in sufficient time to clear it, and so on. It has been an object lesson to the world, and a terrible one. The loss of that huge, magnificent ship – the very pinnacle of all shipping – is in itself great; but the appalling loss of life sinks all else into utter insignificance. The whole horror can only be just lived down, and it will take time.

'I am glad to say that my conscience is absolutely clear. Before leaving the deck at 10 o'clock I did all that human foresight could do, as far as I deemed advisable, with relation to the conditions and vicinity of ice. I tell you this that you may know that we have no worry on that account.

'That there was a great lack of lifeboats is a matter wholly for the Board of Trade, which passes the ship and gives her certificate. The owners and builders are in no way to blame ... We carried more life-saving appliances than were required by the Board of Trade. That the ship was ripped open like a paper bag is for the builders to rectify. The shock was so slight that after coming out in my pyjamas (I had not been asleep) to see why the engines were slowed or stopped, I went back and turned-in.

'I may tell you more by-and-by, but it is all too fresh and comes back too vividly at present; so good-bye for the present.'

Lightoller returned to sea in 1913, as first officer on *Oceanic*. During the First World War, *Oceanic* was commissioned as an armed merchant cruiser on the Northern Patrol, and he became Lieutenant Lightoller of the Royal Navy. Her job was to patrol a 150-mile stretch of water in the area of the Shetland Islands. She ran aground near the island of Foula, and three weeks later *Oceanic* broke up in a storm and was gone.

Lightoller's next assignment was to RMS *Campania,* a Cunard Line ocean liner converted to a seaplane carrier. Lightoller now found himself as the observer in a Short 184 seaplane. In June 1915, during a Grand Fleet exercise off Iceland, he was the observer on the only plane able to get into the air. They located the Blue Fleet, and for the first time in history, a plane sent up by a fleet at sea succeeded in locating an enemy fleet. Just before Christmas 1915 he got his own command, the torpedo boat *HMTB 117*. He and his wife, Sylvia, went to live at 8 East Cliff, Dover, and during his tour with this boat as part of the Dover Patrol on 31 July 1916; Lightoller attacked the Zeppelin L31 with the ship's Hotchkiss guns. For his actions he was awarded the Distinguished Service Cross (DSC) and he was also promoted to commander of the torpedo boat destroyer HMS *Falcon*. On 1 April 1918, Lightoller was lying in his bunk when *Falcon* collided with the trawler *John Fitzgerald*. She stayed afloat for a few hours, eventually sinking just about the same time, and six years to the day as the *Titanic* sinking.

A letter from Lightoller's wife, Sylvia, exists dated 3 April 1918, addressed to Lieutenant C.H. Lightoller DSC on *Falcon*, from 'Cooie' of 8

East Cliff, Dover, with reference to the 1 April 1918 collision and expressing delight at his survival. 'I have you and that's all that matters.'

Lightoller was now given a new command, the destroyer HMS *Garry*. On 19 July 1918, *Garry* rammed and sank the German submarine *UB-110*. The bows of *Garry*, however, were so badly damaged that she had to steam 100 miles in reverse to relieve the strain on the forward bulkheads as she returned to port for repairs. For this action Lightoller was awarded a Bar to his DSC and promoted to lieutenant commander. During the First World War, Sylvia Lightoller became well-known for keeping an open house to Australian and New Zealand Army Corps (ANZAC) troops when they lived at The Cottage in Cockfosters.

At the end of 1918, Lightoller left the Royal Navy as a full commander. On his return to White Star Line he was appointed chief officer of the *Celtic* having been passed over for a position on *Olympic* because the new management wanted to forget *Titanic* and all those associated with her, and it is interesting to note that none of the surviving *Titanic* officers ever got their own commands. However, Lightoller was not interested in remaining chief officer of the *Celtic* indefinitely, so, after well over twenty years of service, he resigned from White Star Line.

Under the title 'Motor Yacht World Tour – *Titanic* Survivor in Command', Sydney's *The World's News* on 27 September 1924 reported:

> Eighty passengers will leave Southampton in October in the *Westward*, of 3,000 tons, the largest motor-sailing yacht afloat, on a tour round the world for scientific research purposes.
>
> The vessel is owned by Commander C.H. Lightoller, who, as second officer of the *Titanic*, remained onboard until she sank.
>
> The *Westward* will call at the West Indies, St Lucia, Barbados, Martinique, St Thomas, Porto Rico, and Jamaica. Thence she will sail through the Panama Canal to the South Sea Islands.
>
> After touching at Malpelo, the *Westward* will make for the Galapagos and the Marquesas Islands. The first of these, composed entirely of extinct volcanic cones, is probably the most weird group of islands in the world.
>
> The Paumotu, or Low Archipelago, famous for its pearling lagoons, will be the next port of call, and then the Society Islands, the Cook Islands, Samoa, Fiji, Tonga, Loyalty and New Caledonia will pass in review before Brisbane is reached.
>
> The vessel will then thread her way up the Great Barrier Reef to Thursday Island, the Gulf of Carpentaria, and through the islands of the Malayan Archipelago, visiting Christmas Island, Cocos, Keeling, Diego Gracia, Elmont, and Male. A complete film of the trip will be taken.

The Navy Department of the United States Hyrographic Office has provided Commander Lightoller with a set of charts covering the tour, and has asked him to report on it. The trip will take approximately ten months.

Commander Lightoller will be accompanied by Mrs [Sylvia] Lightoller, and their eldest son, Roger, who is at present serving as midshipman in the battleship *Valiant*.

The Lightollers opened a guest house and after a few years had some minor success in property speculation. They purchased a discarded Admiralty steam launch in 1929, and had her refitted and lengthened, and converted her into a diesel motor yacht that was christened *Sundowner* by Lightoller's wife. Throughout the 1930s she was used by the Lightoller family mainly for trips around Britain and Europe.

The Manchester Guardian in July 1932 reported under the heading '*Titanic*'s Second Officer – A Mysterious Report'

Commander C.H. Lightoller, who was second officer on *Titanic*, had a shock on the twentieth anniversary of the sinking of the ship, with the loss of more than 1,600 lives, after colliding with an iceberg on her maiden voyage. The following cable was received in London from San Francisco.

While the anniversary of the sinking of the ill-fated *Titanic* is being generally observed, the ill-fated liner's second mate, Charles Herbert Lightoller, whose bravery was one of the redeeming features of the disaster, is lying forgotten and destitute in Santa Rosa Hospital near San Francisco.

The real facts are that Commander Lightoller is in the best of health – and spirits – at his home in Putney, and the only explanation he can offer of the report is that the man in hospital is suffering from delusions. Commander Lightoller, who retired from the sea shortly after the war [First World War], has not been to San Francisco for many years.

It was in the following year that newspapers reported that he intended to go back to St Paul's island in a quest to find treasure that he believed was buried there. In 1935 he published *The* Titanic *and Other Ships*, and on 1 November 1936 he recited his account for the BBC radio programme I *Was There*.

In July 1939, Lightoller and his wife, Sylvia, were asked by the Royal Navy to perform a survey of the German coastline. This they did under the guise of an elderly couple on vacation in their yacht. When the Second World War broke out in the following month, the Lightollers were raising chickens in Hertfordshire. Their youngest son, Brian, was a pilot in the RAF, and on the first night of hostilities he was killed in a bombing raid on Wilhelmshaven. On 31 May 1940, Lightoller received a telephone call from the Admiralty asking him to take the *Sundowner* to Ramsgate, where a navy crew would take over and sail her to Dunkirk. Lightoller informed them that nobody would take the *Sundowner* to Dunkirk but him.

On the 1 June 1940, 66-year-old Lightoller, accompanied by his eldest son, Roger, and an 18-year-old sea scout named Gerald Edward Ashcroft, sailed *Sundowner* out of Ramsgate headed for Dunkirk and the trapped British Expeditionary Force. Although the *Sundowner* had never carried more than 21 persons before, they succeeded in carrying a total of 130 men from the beaches of Dunkirk. In addition to the 3 crew members, there were 2 crew members who had been rescued from the motor cruiser *Westerly*, 3 naval ratings rescued from waters off Dunkirk, plus 122 troops taken from the destroyer HMS *Worcester*. Despite numerous bombing and strafing runs by Luftwaffe aircraft, they all arrived safely back in Ramsgate just about twelve hours after they had departed. It is said that when one of the soldiers heard that the captain had been on the *Titanic*, he was tempted to jump overboard. However his mate was quick to reply that if Lightoller could survive the *Titanic*, he could survive anything and that was all the more reason to stay.

Commander Lightoller joined the Home Guard, but the Royal Navy engaged him to work with the Small Vessel Pool until the end of the Second World War. His son, Roger, joined the Royal Navy and commanded motor gun boats. He was killed in an enemy commando raid on Granville on the northern coast of France during the latter stages of the conflict.

After the Second World War, Lightoller went into the boat building business, first with an outside partner, then with his son, Trevor. *Sundowner* was retrieved and refurbished, and Richmond Slipways, located at 1 Duck's Walk near Richmond Bridge, East Twickenham, specialized in police river launches. Lightoller and his wife, Sylvia, lived over their place of work, and he and David Blair kept in touch.

Charles Lightoller died of heart disease on 8 December 1952, aged 78. He was cremated at Mortlake Crematorium and the ashes scattered in the Garden of Remembrance. It is possible that he may have succumbed prematurely to his illness. A life-long pipe smoker and suffering from heart disease, he was living in London during the city's great smog of 1952 when he died from complications of his illness.

Under the title 'Wartime Hostess on Visit' *The Canberra Times* on 18 October 1962, carried the following report:

> A woman known to hundreds of Australian servicemen for her 'open houses' in wartime London, is to return to Australia for the first time in 60 years.

She is Mrs Sylvia Lightoller, 77, who was born Sylvia Hawley-Wilson at Hell's Hole, a gold mining settlement near Blayney, N.S.W. Mrs Lightoller will spend six months in Australia and New Zealand, seeing as many of the 'boys' as she can. She will visit Sydney, Melbourne and Brisbane, then Auckland, Wellington and Christchurch.

Mrs Lightoller is the widow of Commander Charles Herbert Lightoller, the only senior officer to survive the sinking of the *Titanic* on her maiden voyage in 1912. He died in 1952, aged 78.

She came to London for a year's schooling and met her future husband on a ship going home. They were married at St James's Church, Sydney, and came back to London when Mrs Lightoller was 17 years old. (She has not seen Australia since.)

Mrs Lightoller kept open house at The Cottage, Cockfosters, London, for Australian and New Zealand servicemen during the war. She told A.A.P: 'I am looking forward to seeing all the changes in Australia ... Why, it was still in the horse and buggy days when I left. But most of all I am looking forward to seeing the "boys" and their families,' she added. Mrs Lightoller is due in Sydney by air on 25 October.

Louise Patten, the financier and author, is Lightoller's granddaughter. Charles Lightoller was depicted by a 34-year-old Londoner named Jonathan 'Jonny' Mark Phillips in the 1997 film *Titanic*.

***

Herbert 'Bert' John Pitman was born on 20 November 1877 at Sutton Montis in the birth district of Wincanton, Somerset. He was the son of a farmer named Henry Pitman (1813–1880) and Sarah A. Pitman – née Marchant (born 1851). His father died when he was aged 3, and his mother remarried a man named Charles Candy.

At the time of the 1881 census he was living on a 112-acre farm on Sutton Road in Sutton Montis, Wincanton, with his widowed mother, older brother, William Henry, and younger sister, Ida Mary. In 1891 he lived at Rimpton, Sherborne, with his mother – now known as Sarah Candy – and his step-father, Albert Charles. He is also known to have lived at Castle Cary. Five players from the Castle Cary Cricket Club played in the team that became the only ever Olympic cricket champions when they won in Paris in 1900, and Herbert Pitman became a cricket enthusiast, probably for Somerset County Cricket Club, although the county have never won the County Championship.

He received the shore part of his nautical training in the navigation department of the Merchant Venturers' Technical College in Bristol. He first went to sea in 1895, and part of the United States Senate inquiry summed up his early career:

Senator Smith: 'How long have you been engaged in marine employment?'

Pitman: 'About 17 years – four years with James Nourse (Limited); three years as an officer in the same employ; about twelve months in the Blue Anchor Line, running to Australia; six months in the Shire Line, running to Japan; and five years with the White Star Line.'

Senator Smith: 'In what capacity did you serve with the White Star Line?'

Pitman: 'Second, third and fourth officer; second officer for two months.'

Senator Smith: 'On what vessels of the White Star Line did you serve?'

Pitman: 'On the *Dolphin*, the *Majestic*, and the *Oceanic*.'

In his private life he became a member of the Hatfield Abbey Lodge of Freemasons in 1909, remaining so until he died, and he was also keen on stamp collecting all his life.

While reporting on the British inquiry the *Daily Herald* stated:

It was an officer who again took up the tragic tale of *Titanic*. He was the third officer, Mr Herbert John Pitman. He looked older in years than Mr Lightoller; his dark close-cut hair flecked with grey. He was saved in [life]boat number 5.

During the First World War – 1914 to 1918 – Pitman served aboard the White Star Liner *Teutonic*, which was converted into a troopship. In 1916, he was commissioned as lieutenant in the Royal Naval Reserve, and saw out the conflict as a stores officer on a destroyer. He ended the war as a lieutenant commander.

Less than one year after *Titanic* sank, Pitman failed his vision test and became a purser, and on returning to service with the White Star Line, in 1921, he is listed as a purser aboard *Adriatic* and *Olympic*.

At Paddington in London in June 1922, he married a New Zealand girl named Mildred 'Mimi' Kalman (1886–1933), and the *Auckland Evening Post* on 30 June 1922, reported:

A wedding of New Zealand interest took place in London, when Miss Mimi Kalman, youngest daughter of the late Mr Charles Kalman and Mrs Kalman, of Park Avenue, Auckland, was married to Mr Herbert

John Pitman, of Castle Cary, Somerset. Miss Kalman was well-known in musical circles in Auckland and Wellington.

Mimi Pitman died on 20 October 1933, aged 47.

During the Second World War, Lieutenant Commander Pitman served as purser aboard SS *Mataroa*, where his duties were to meet the needs of troops being transported on the ship. *Mataroa* was involved in a few military actions.

Pitman retired in spring 1946, after spending over fifty years in maritime service. He spent his retirement living with his niece in the village of Pitcombe near Wincanton in Somerset. He was created a Member of the Order of the British Empire (MBE) in 1948.

On 27 August 1954, several Somerset newspapers published an article based on an interview with Pitman, which contained some interesting anecdotes concerning his life. He attended the premiere of A *Night to Remember* in London in 1958, and was photographed with technical advisor (and survivor) Joseph Boxhall and producer William MacQuitty, who had watched *Titanic*'s launch in May 1911.

Pitman and Boxhall kept in touch throughout their lives. Both men admitted to bearing the burden of a bad conscience for not having tried to rescue anyone from the water that dreadful night.

While on holiday in Bournemouth on 18/19 March 1961, Pitman, along with his nephew and his wife, paid a surprise visit to the home of Joseph Boxhall in Christchurch.

Pitman died on 7 December 1961, aged 84, of subarachnoid haemorrhage (bleeding on the brain); almost nine years to the day after Charles Lightoller. He was buried at St Leonard's Parish Churchyard in Pitcombe. His death left Joseph Boxhall as the only survivor of the *Titanic* officers.

He is the only officer to have his photograph taken in colour. A special maritime memorabilia auction including his 'Safe – Bert' telegram was held in London in 1991, another auction, which included his ephemera, was held in Southampton in 1998, and in 2016 his great niece, who had watched cricket with him, brought three items associated with him to the BBC programme Antiques Road Show, recorded in Chippenham, including a facsimile of his discharge book.

***

Joseph 'Joe' Groves Boxhall Jr was born into a seafaring family on 23 March 1884 at Sulcoates, Kingston-upon-Hull, East Yorkshire. Having an older sister, he was the second child of Captain Joseph Boxhall (1858–1928), who was a respected master with the Wilson Line of Hull, and his wife, Miriam 'Mirie' Mary Abigail – née Groves (born 1861). They had married in Hull on 25 June 1881. His uncle, Charles, was an official at the Board of Trade. When he was aged about 11, in 1896, his father took him and his elder sister on his steamer SS *Alecto*, to the Chelsea district of Boston in the United States, where he 'enjoyed July 4th celebrations.'

Despite suffering with seasickness as a boy, Boxhall followed family tradition when he travelled to Liverpool on 2 June 1899, to join the barque *Cambrian Warrior* of the William Thomas Line. He travelled extensively during his four year apprenticeship, and then he returned to Hull to work with his father on the Wilson Line. His diary written between 1899 and 1902 was discovered in 2003 among the archives at Hull Trinity House, and in an article based on the information in the diary the local newspaper described him as an 'ambitious seaman'.

The family home was at tree-lined 27 Westbourne Avenue, and he studied at Trinity House in Hull. He obtained his extra master's certificate in 1907, and joined the White Star Line. He was confirmed as sub-lieutenant with the Royal Naval Reserve on 1 October 1911, serving with *Arabic* and *Oceanic II*, before he was transferred to *Titanic*.

The Adelaide Observer on 9 November 1912 stated:

Memories of the loss of the *Titanic* were revived on Saturday by the arrival of Mrs [Evelyn] James [1883 –1938], daughter of Mr and Mrs W.H. Marsden, of Hoyleton, South Australia, by the White Star steamer *Irishman*. Mrs James (then Miss Marsden) was among the stewardesses saved from the ill-fated vessel.

She was recently married to Dr James, who is a surgeon on the *Irishman*. Another *Titanic* survivor, Mr M.J. [sic] Boxhall, is second officer on the same vessel. He held the position of fourth officer on board the mammoth liner.

During the First World War, on 27 May 1915, Boxhall was promoted to lieutenant in the Royal Naval Reserve and saw service for a year aboard the battleship HMS *Commonwealth*. He then saw service in Gibraltar, where he commanded a torpedo boat.

After coming out of an engagement to an Australian girl, on 25 March 1919, he married a Sheffield girl named Marjorie Beddells (1882–1972), the daughter of an industrialist, at St Andrew's Church in the Sharrow district of her home town (now a city). The marriage was known to be a happy one. They did not have any children.

He returned to service with the White Star Line two months after his marriage, sailing three more times to New York; twice on *Cedric* in 1919, and on SS *Regina* in 1923. He was appointed lieutenant commander in the Royal Naval Reserve on 27 May 1923, and he signed on as second officer with *Olympic* on 30 June 1926. After the White Star-Cunard merger in 1933, he served as a senior officer on RMS *Aquitania*. He retired from the sea in 1940, his last voyage being as chief officer of RMS *Scythia*.

In 1937, he featured prominently in a Cunard Line advertisement entitled 'On Watch', which appeared in the popular periodicals *Life* and *National Geographic*. He acted as technical adviser for the film A *Night to Remember* which was released in 1958, and he took part in a BBC radio broadcast on 22 October 1962. At the time he was living at 11 Walcott Avenue in Fairmile, Christchurch, Hampshire.

On 18/19 March 1961, he received a surprise visit from Herbert Pitman, who was on holiday in Bournemouth, along with his nephew and his wife, less than a year before Pitman died, and in the following month he wrote:

> I am now a very old man of 77, and most of my time [is spent] reading and dozing off to sleep ...

His health began to deteriorate to the point where he had to be admitted into Christ Church Hospital, and he died of cerebral thrombosis on 25 April 1967, aged 83. He was cremated and, following his wishes, his ashes were scattered at sea by the Cunard Line passenger liner RMS *Scotia*, in the area he had calculated to be *Titanic*'s final resting place. Marjorie Pitman died in Bournemouth in 1972. In 2006, the last surviving officer of *Titanic* was commemorated with a green plaque at his former home at 27 Westbourne Avenue, Kingston-upon-Hull.

As a point of interest, green plaques on the same avenue are dedicated to the crime writer, Dorothy L. Sayers (1893–1957), who had lived in the avenue from 1916 to 1917; the actor Ian Carmichael (1920–2010) – who portrayed her famous creation, Lord Peter Wimsey on television and radio – lived on the avenue as a boy; as did Gerald Thomas (1920–1993), and Ralph Philip Thomas (1915–2001), who produced the 'Carry on' and 'Doctor' film comedies respectively.

***

Harold Godfrey Lowe was born on 21 November 1882 in his grandfather's home 'Bryn Lupus' (Wolf Hill) at Eglwys Rhos, Llanrhos (Church on the Moor) in Caernarvonshire (now Conwy County). He was the second son of five (the last two being twins), in a family of eight children born between 1878 and 1894, to George Edward Lowe (1848–1928), and his wife, Emma Harriette – née Quick (1856–1909). They met in Liverpool, and were married on 6 June 1877. George Lowe grew up in Chester, where his family ran Lowe and Company – silversmiths, goldsmiths, jewellers and watchmakers – with other branches in Liverpool, Llandudno and Barmouth.

The year after Harold Lowe was born the family moved 40 miles south of Llanrhos, to the county of Merionethshire (now Gwynedd), and the 1891 census has them living at the Castle Hotel in Llandanwg; his father was recorded as a landscape and cattle painter, and his mother worked as the hotel manageress. In 1893 they moved to Barmouth, still in Merionethshire, where they lived in a house called 'Penrallt'.

Being a town near the sea, the Lowe boys became familiar with sailing and boat-handling. However, on 27 December 1895, when Harold Lowe was aged 13, his eldest brother, George, aged 17, drowned in a boating accident. His body was found in Aberamffra Bay. On 14 September of the following year, Harold Lowe almost suffered the same fate. He was out in his father's punt at Aberamffra Bay when the weather turned bad and the boat capsized; but Lowe managed to swim back to the shore.

However, Lowe showed confidence in his seamanship at an early age. Barmouth residents recalled to the *Titanic* enthusiast, Inger Shiel, an incident in which he and some of his friends took a small boat so far out to sea that those ashore feared they were in danger and sent out a lifeboat to rescue them. As it drew alongside them Harold Lowe cooly asked them where they were bound for.

Harold Lowe was educated with his younger brother, Edgar, at the Barmouth Board School, and then at the newer Barmouth County Intermediate School, and at the age of 14 his parents wanted him to start an apprenticeship doing some kind of office-based work with a successful businessman in Liverpool, but he decided that,

> ... [he] was not going to work for anybody for nothing ... I wanted to be paid for my labour.

He ran away and worked as a ships' boy aboard seven Welsh coastal schooners as he studied to attain his maritime certification. He joined the Royal Naval Reserve in about 1904, being given the service number 13213. On 10 November 1904, he was serving aboard SS *Prometheus* of the Alfred Holt Line when it was involved in a collision with a larger ship named HMS *Picton* in the harbour at Amsterdam, and he gained his first experience of being a witness at an inquiry. In 1906 he served on the steam ships SS *Justin*, SS *Fabian*, SS *Ardeola* and SS *Charma*, and after on failure to gain his second mate's qualification, it was finally issued on 23 August 1906, which made the local news. His first mate's certificate was issued on 27 July 1908. He served on the steamships SS *Madeira* and SS *Oron* that year. He served aboard the steam ship *Addah* on 1909, and gained his master's certificate on 12 November of that year; a document which he had framed.

He worked on board SS *Zaria* on the West Africa service until 18 March 1911, before he joined the White Star Line on 8 April 1911. At that time he lived in the household of Isaac Arthur Jones. His employments were as third officer aboard SS *Tropic*, SS *Haverford*, SS *Mersey* and SS *Belgic*, for voyages to Australia; he was chief officer on *Mersey*; although some discrepancies have been suggested concerning this period of his service record.

On his return home to Barmouth after the disaster 1,300 people attended a reception in his honour, held at the Picture Pavilion, and he was presented with a commemorative gold watch and chain bearing the inscription:

> Presented to Harold Godfrey Lowe, 5th Officer, RMS *Titanic*, by his friends in Barmouth and elsewhere, in recognition and appreciation of his gallant services at the foundering of the *Titanic*, 15 April 1912.

It was reported:

> Mr Lowe was so overcome by his feelings at this mark of esteem on the part of his Barmouth friends that he was only able to express his feelings in a single sentence.

By September 1912, Lowe was in Australia as third officer on *Medic* where it was reported:

> For various reasons, one of which is that he is 'so sick of it all' and desires, if possible, to forget about the past. Mr Lowe is disinclined to discuss the tragedy.

At St Paul's Church in Colwyn Bay, on 24 September 1913, Lowe married Ellen 'Nellie' Marian – née Whitehouse (1885–1947). They had two children, Florence Josephine 'Josie' Edge (1914–1999) and Harold William George (1916–1999).

During the First World War, Lowe was a commander in the Royal Naval Reserve, and was promoted to lieutenant on 20 July 1915. For most of the conflict he was assigned to HMS *Donegal*, and he saw service in Vladivostok during the Russian Revolution and Civil War of 1919; attaining the rank of lieutenant of the Royal Naval Reserve. After the First World War, he returned to service with the International Mercantile Marine and the White Star Line, serving on numerous steamships of the line.

On 4 July 1919, Lowe was involved in another boating accident. He was out in Barmouth harbour with a friend and a young lad named

Robert Lunt. The boat capsized, and as the small rescue vessel belonging to Harry Jones went out to help them, Lowe held on to the upturned boat with one hand and held the lad above the water with the other. They were all brought to shore, apparently, 'none the worse for the experience'.

He was initiated as a Freemason at the St Trillo Lodge on 13 May 1921, and he received the Reserve Decoration (RD) medal in August 1927, which was awarded to officers who have achieved at least fifteen years' service. It was last awarded in 1999. He was released from the Royal Naval Reserve as lieutenant commander on 21 November 1927. His last ship was SS *Doric*, serving on her as second officer from 27 July 1928 to 13 November 1930.

In 1931 he resigned from White Star Line, and lived in a large house at 1 Marine Crescent on the waterfront overlooking the River Conway in Deganwy, where he came to be known as '*Titanic* Lowe'. He served as a 'high-profile' Conway town councillor for two terms between 1932 and 1938. He also volunteered as a church warden.

In 1936, a young girl named Annette Roberts visited the Lowes for afternoon tea. She had been instructed to call him commander, but found him to be a 'jovial man'. She recalled 'the huge table set for tea, with a dainty brass shovel and brush for clearing away the crumbs. Apparently, Commander Lowe thought it was hugely funny to secrete a whoopee cushion under the covering of Annette Robert's chair at the table, and when she sat down she was 'ashamed and embarrassed when a loud noise emanated from beneath her.'

> I remember Mr Lowe as full of fun, there was a lot of laughter and jokes thrown around that afternoon. I only listened to the conversation because children really did only speak when spoken to in those days. I remember it was a very dainty afternoon tea with sandwiches and scones, it was quite grand. The *Titanic* wasn't mentioned that afternoon, but I do know Mr Lowe had been a very brave man that night.

On 19 November 1937, he was involved in another boating accident. He and a local railway clerk named William Parry had set out in a dinghy towards his motor launch in the River Conway in Deganwy. As they reached the launch, Lowe fell into the sea between the two vessels and his knee boots were soon waterlogged. Fortunately, Parry grabbed hold of his collar and kept him afloat, and eventually managed to pull his exhausted companion into the launch. The unfortunate incident, embarrassing for an old sea dog, was reported in the local newspaper, probably against Lowe's wishes.

During the Second World War, Lowe volunteered his home as a sector post and served as an air raid warden until ill health forced him into a wheelchair.

After suffering a stroke, Lowe became the first of the surviving officers to die, dying of hypertension in Deganwy on 12 May 1944, aged 61, and he was buried in Llandrillo-Yn-Rhos Churchyard at Rhos-on-Sea, Colwyn Bay. Nellie Lowe died on 10 February 1947, aged 63, and was buried with her husband.

On 22 April 2012, a service of commemoration was held at the church, attended by his great nephew, John Harold Lowe, who lived in his old house at Deganwy, and several local dignitaries. Due to the determination of a local schoolgirl a slate plaque has been placed in the harbour master's office at Barmouth, and a blue commemoration plaque at his former home in Deganwy.

John stated that:

He [Harold Lowe] was very reticent to talk about it. He talked to his son, my uncle about it, but I was forbidden to talk about it.

Rene Harris, who was rescued in collapsible B, described Officer Lowe as 'One of the finest men it has been my privilege to meet.' She presented him with a number of nautical items to thank him for his gallantry on the 'Night to Forget' as she put it.

***

James 'Jim' Paul Moody was born on 21 August 1887 at 17 Granville Road, South Cliff in the popular holiday resort of Scarborough, North Yorkshire. He was baptised at the Anglican church of St Martin-on-the-Hill in Scarborough. He was the youngest child in the family of three boys and a girl, all born in Scarborough between 1881 and 1887, to a Scarborough-born solicitor and councillor named John Henry Moody (born 1857) and his wife, Evelyn Louisa – née Lammin (1858–1898), who came from Fulham.

Sir James Harland of Harland and Wolff was also born in Scarborough.

According to the 1881 census the Moodys still lived at Granville Road, where they employed a nurse, a cook and a housemaid. Moody's mother died in 1898 and his father remarried a woman named Annie, and had four more boys. Moody received a good education at the Rosebery House School in Scarborough.

He was on the Royal Navy training ship *Conway* in the River Mersey as a cadet in 1902 to 1903, which counted as one year's sea time. He joined the William Thomas Line's ship *Boadicea* as an apprentice, and during a trip to New York, weather conditions were so horrific that it drove one of his fellow apprentices to commit to suicide. On 29 May 1904, Moody witnessed a massive fire that broke out in the shore freight yards in New Jersey, which destroyed six piers and dozens of barges. He then sailed with his ship to Sydney, Australia.

He went to Liverpool and stayed at the Sailors' Home so he could apply for his second mate's certificate, which he passed on the first examination on 24 June 1907. At the time his address was recorded as St James House in Grimsby. His first qualification was issued at the port of Grimsby on 1 July 1907. He was described as being 5 feet 11 inches tall, with a fair complexion, light brown hair and blue eyes.

He went into steamships and worked on cargo and oil tankers, before gaining his first mate's certificate. He attended the King Edward VII Nautical School for a short while in 1910, which prepared officers for their Board of Trade examinations, and he attained his ordinary master's certificate on 26 April 1911.

He joined the White Star Line as sixth officer on 17 August 1911, and served aboard *Oceanic* as her fifth officer from 30 December 1911 to 26 March 1912. At that time his shore address was the house of his uncle in Grimsby, where it is suggested he had a romantic interest that he told his brother, Christopher, 'might be the one.'

Soon after his death the White Star Line sent a letter to Moody's brother, Christopher, requesting a £20 deposit to return his body to England, and stating that all further costs would have to be covered by the Moody family or his body would be buried with other victims of the disaster at Halifax in Nova Scotia, Canada. In reality his body was never identified.

In April 1912, a requiem service for the 'repose of his soul' was held at St Augustine's Church in Grimsby, and the Moody family donated an altar set dedicated to the memory of James Moody who remained 'steadfast unto death' which is preserved in the church. The wardens of the Church of St Martin-on-the-Hill in Scarborough erected a memorial plaque in the church dedicated to him on 7 August 1913, which bears the words: 'Be thou faithful unto death, and I will give thee a crown of life.' In 2002 a memorial and brass plaque was erected by his descendants in the Scarborough Lifeboat House on the sea front. A blue plaque was unveiled by the Scarborough and District Civic Society in 2012 at his birthplace at South Cliff, which was attended by some of his descendants, and he is commemorated on the headstone of his mother's grave at the Manor Road Cemetery in Scarborough, which states:

> Also in loving memory of James Paul Moody, her youngest son, born August 21st 1887. Gave up his life in the wreck of the SS [*sic*] *Titanic*, April 15th 1912. Greater love hath no man than this, that a man who lay down his life for his brothers.

Chapter 9

DID CAPTAIN SMITH SURVIVE THE SINKING?

There have been various accounts of Captain Smith's last moments concerning the disaster, and it is almost certain that if his distinctive body and uniform was found it would have been identified, which it was not.

Harold Bride was the junior wireless officer on the ship, and part of his testimony to the United States Senate inquiry when question by Senator William Alden Smith went as follows:

Senator Smith: 'When did you last see the captain [Smith]? When he told you to take care of yourself?'

Bride: 'The last I saw of the captain [Smith] he went overboard from the bridge, sir.'

Senator Smith: 'Did you see the *Titanic* sink?'

Bride: 'Yes, sir.'

Senator Smith: 'And the captain [Smith] was at that time on the bridge?'

Bride: 'No, sir.'

Senator Smith: 'What do you mean by overboard?'

Bride: 'He jumped overboard from the bridge when we were launching the collapsible lifeboat.'

Senator Smith: 'I should judge from what you have said that this was about three or four minutes before the boat [*Titanic*] sank.'

Bride: 'Yes. It would be just about five minutes before the boat [*Titanic*] sank.'

Senator Smith: 'About five minutes?'

Bride: 'Yes.'

Senator Smith: 'Do you know whether the captain [Smith] had a life belt on?'

Bride: 'He [Smith] had not when I last saw him.'

Senator Smith: 'He [Smith] had not?'

Bride: 'No, sir.'

Senator Smith: 'Did the bridge go under water at about the same time?'

Bride: 'Yes, sir. The whole of the ship was practically under water to the forward funnel, and when I saw her go down the stern came out of the water and she slid down fore and aft.'

Senator Smith: 'The captain [Smith] at no time went over until the vessel sank?'

Bride: 'No, sir.'

Senator Smith: 'He [Smith] went with the vessel?

Bride: 'Practically speaking; yes, sir.'

At least three eyewitnesses – a second class passenger named Charles Eugene Williams (1888–1935), who was in lifeboat 14, and two crew members, Fireman Harry Senior (1881–1937) and Walter 'Wally' Hurst (1888–1964), both in collapsible lifeboat B – say they saw Captain Smith in the water as the lifeboats were trying to get away from the doomed liner, and Williams said he actually had a short conversation with him. Can all three of them be wrong?

A few months later Captain Peter Pryal of Baltimore, who had sailed with Captain Smith, was certain he met him walking the streets of that city.

The usual explanation is that it was simply a case of mistaken identity, but is there any possibility that it could have really been Captain Smith?

The *Daily Sketch* on 30 April 1912 reported, under the heading: 'His Last Act Was to Save a Child's Life – Refused to get into a [life]Boat.'

> Of all the wild and irresponsible messages that were sent to this country in the first hours following the sinking of the *Titanic* the one that caused the grief to Englishmen was the statement that Captain Smith had committed suicide on the bridge of his ship. That statement was quickly contradicted. It was proved that Captain Smith died like a sailor, but the exact manner of his death was not described. Many eyewitnesses have testified to seeing the captain on the bridge as the great liner was engulfed, and others say they saw the officer dive from the bridge just before the ship sank, but according to an interview published in the *New York World* Captain Smith died the greatest of all deaths. His last act was to save the life of a child.
>
> The story is told by Charles Eugene Williams, coach of the Harrow Racquet Club, who was one of those saved from the *Titanic*. Mr Williams is the guest of Mr George E. Standing, and the latter gave the *New York World* reporter the account as told to him by Mr Williams.
>
> 'He [Charles Williams] is a good swimmer,' said Mr Standing, 'and went overboard with a life preserver when he couldn't stay on deck any longer. He was in the icy water for over two hours before he was finally hauled into one of the lifeboats.' He says that he saw Captain Smith swimming around in the icy water with an infant in his arms and a lifebelt. When the small [life]boat went to his rescue Captain Smith handed them the child, but refused to get in himself.
>
> He did ask what had become of First Officer Murdoch. We told him Murdoch had blown his brains out with a revolver [which was incorrect]. Then Captain Smith pushed himself away from the lifeboat, threw his lifebelt from him and slowly sank from sight. He did not come to the surface again.
>
> It may be stated that there is no official confirmation that Mr Murdoch shot himself: we give the whole account as it appeared in the *New York World*.

Most of Williams's story is borne out by Harry Senior, except that Senior witnessed Captain Smith swim back to the ship after he had rescued the baby. Senior arrived on SS *Lapland*, and in interviews for *The New York Times* on 19 April 1912 and the *Illustrated London News* on 4 May 1912, he stated:

'The ship [*Titanic*] was pretty near sinking then, and the captain [Smith] shouted, "Each man for himself". I had noticed him on the bridge before that. He was pacing up and down sending up rockets and giving orders. It is a dirty lie to say that such a man as he shot himself.'

When the Captain's order came Senior dived over the side.

'As I was swimming to the [life]boat I saw the captain in the water. He was swimming with a baby in his arms, raising it out of the water as he swam on his back. He swam to a [life]boat, put the baby in, and then swam back to the ship. I also had picked up a baby, but it died from the cold before I could reach the [life]boat.'

Walter Hurst and his father-in-law, William Mintram (1866–1912) – Hurst was married to Mintram's daughter, Rosina May (1887–1969) – served as firemen on the ship. They met each other shortly before *Titanic* went down. Mintram had a life jacket and he gave it to his son-in-law. This act of courage probably cost Mintram his life, and may well have contributed to the fact that Hurst was able to stay on the upturned collapsible lifeboat B and survive.

Hurst's daughter, Rosina Ellen Hurst – later Broadbere (1916–2002), related:

My father was one of the last ones off and he was in the collapsible [life] boat that was upside down in the water, he was on the top of that, and he always said that, you know the captain [Smith] of the ship, they say all sorts of things about what happened to him, well he said somebody swam up to the [life]boat and he couldn't get on because there was so many on there, and he said 'Good luck, boys' and he went. My dad [Hurst] swore it was the captain that said that.

It is known that Fifth Officer Lowe rescued at least four men from the water when he returned to the area of the disaster, and one of them died. One of the other three may have been an Italian named Emilio Portaluppi. However, there is stronger evidence to suggest that Portaluppi was rescued in lifeboat 4. If that is so, who was this fourth man? Others, such as Emily Ryerson, suggested they picked up as many as six or seven men.

Captain Smith must have been in the icy water for quite a long time, so would he have been in any conscious state to say who he was, and would he have been recognisable – especially in the dark? Fifth Officer Lowe and Thomas Threlfall were in the lifeboat, but Lowe was preoccupied organising the rescue attempt and Threlfall was busy looking after one of the female passengers.

It is believed that a survivor was picked up by *Carpathia* and seemingly rushed away from the eyes of those around him. Who was this man? Some suggest it was Bruce Ismay. However, at least one first class passenger, Jack Thayer, was allowed to go into his cabin and speak to him, so what would have been the point of getting him away from the others and then let him be seen? It is also known that the survivors of the crew were not disembarked with the passengers, but were taken by a tender, the *George E. Starr*, at Pier 61. There seems to have been an attempt to keep the crew away from awkward questions until they had been briefed. The root of all evil also raised its head. According to the *London Daily News*, one of the things the Senate inquiry intended to find out was:

> Why a Marconi Company's official sent a wireless to the operator on *Carpathia* on Thursday, stating: 'Say nothing. Hold your story for dollars in four figures.'

If it was Captain Smith on *Carpathia* surely the people who were rescued in the same lifeboat as him would have recognised him and remembered? However, it was too dark to distinguish faces, and in fact, first class passenger Jane Anne Hoyt (1881–1932), who was in collapsible lifeboat D, stated later that she did not even realise until they reached *Carpathia* that her husband, Frederick Maxfield Hoyt (1873–1940), had jumped into the water as *Titanic* went down and had been dragged into the same lifeboat as her shortly afterwards.

On 23 July 1912, numerous newspapers around the world reported: '*Titanic* Mystery' – 'Captain Smith said to be in Hiding':

> Peter Pryal, veteran captain of Montreal, and a long-time friend of Captain Smith, of the *Titanic*, swears that Captain Smith was not drowned. Pryal declares that he saw him recently in Baltimore. Captain Smith tried to evade him by taking the train bound for Washington [DC]. It is stated that Captain Smith's nephew, a resident of Baltimore, disappeared the day after his uncle's re-appearance.
>
> Captain Peter Pryal (says the New York correspondent of *The Daily Telegraph*, on July 21 [1912]), one of the oldest mariners of Baltimore, well-known in shipping circles, who sailed with Captain Smith of the *Titanic* when the latter was commander of the *Majestic*, made a startling statement yesterday. He had seen and talked to Captain Smith on Friday at Baltimore and St Paul Streets.
>
> He declares that he walked up to Captain Smith and said, 'Captain Smith, how are you?' Then, according to Captain Pryal, the man answered, 'Very well, Pryal; but please don't detain me. I'm on business.' Captain Pryal says he followed the man, and saw him buy a ticket for Washington [DC], and as he passed through the gate of the railroad

station he turned, recognised Pryal again, and remarked, 'Be proud, shipmate, until we meet again.'

'There's no possibility of my being mistaken,' said Captain Pryal yesterday. 'I have known Captain Smith too long. I know him even without his beard. I firmly believe that he was saved, and in some mysterious manner brought to the United States. I am willing to swear to my statement.'

Captain Pryal is a plain citizen, a teetotaller and a churchgoer, and one of the last men, it is believed, to desire publicity. He realises the incredulity his statement will awaken. 'People will say I'm either a second-sight man or insane: but they are mistaken. I simply believe the evidence of my eyes and ears.'

Dr [Matier] Warfield of Baltimore [said to be related to the famous Mrs Wallace Simpson] vouches that Captain Pryal is 'perfectly sane.'

Captain Smith had a distinctive look, and Captain Pryal actually spoke to the man. Could it really have been a case of misidentification of both visual and voice?

However, the White Star Line management dismissed Captain Pryal's report out of hand, and it was reported on 1 August 1912:

The story told by Captain Peter Pryal, of Baltimore, to the effect that he met Captain Smith, the commander of the *Titanic*, who was supposed to have gone down with his vessel, on the streets of Baltimore on Friday last, and talked with him was received with incredulity at the offices of the White Star Line here.

'We have heard nothing to indicate that Captain Smith did not meet death when the *Titanic* went down,' said the manager. 'This Baltimore captain's story must be either the result of delusion or mistaken identity.'

Captain Pryal had stated that he sailed with Captain Smith when the latter commanded the *Majestic*; that he met him on the street in Baltimore on Friday, talked with him and watched him depart for Washington DC.

The Baltimore newspapers reported on 20 July 1912:

Captain E.J. Smith, commander of the ill-fated *Titanic*, was not drowned when his ship went down in the mid-Atlantic some months ago, but escaped and in some manner reached land, and is now in this country, having been seen and conversed with here yesterday. Is the statement of Peter Pryal, an old mariner, who was quartermaster on the SS *Majestic* of the White Star Line 30 years ago when Smith was its captain. Pryal declares he saw Captain Smith first last Wednesday, and again yesterday when he talked with him. He says that when he saw the captain on Wednesday he was dressed in a neat business suit, carrying two suitcases and appeared unconscious of his surroundings, failing to reply when spoken to by Pryal. Returning to the same place yesterday, Pryal declares Captain Smith again appeared after some time had elapsed, and as he approached Pryal greeted him with: 'Captain Smith, how are you?' Very well, Pryal, but please don't detain me, I am on business', was the reply, says Pryal.

Pryal followed his old commander who turned, and seeing him made an unsuccessful attempt to shake off his pursuer. Pryal followed him to the B & O Depot, where Smith purchased a ticket for Washington [DC], and as he passed through the train gate he turned and with a smile, said, 'Be good to yourself until we meet again.'

Pryal declares he will take an oath to the incident, and is confident that Captain Smith was the man he talked with, as he would know him anywhere even without his whiskers.

A 'Special Dispatch' to *The Baltimore Enquirer* stated – Baltimore, Maryland, 20 July 1912:

The statement that Captain Smith, commander of the ill-fated *Titanic*, was not drowned in the disaster, but was seen safe-and-sound Friday morning in Baltimore, was made today by a retired mariner, who claims to have been a shipmate of Captain Smith for more than 17 years.

Peter Pryal, 9074 Valley Street, who was a quartermaster on the steamship *Majestic*, of the White Star Line, 30 years ago, when Captain Smith commanded the vessel, made the statement, and added that he had not only seen the captain, but talked with him.

Mr Pryal also said that he saw Captain Smith last Wednesday morning, but was sceptical as to his identity, and to confirm his belief that the captain was alive, went to the same spot Friday morning to see the captain again. So shocked was Mr Pryal at seeing the man he believed dead that on his return home he suffered a nervous breakdown.

At 9 o'clock Friday morning he went to Baltimore and St Paul Streets and stood on the corner for almost an hour. Finally to his astonishment he saw the same man approaching him. Walking up to him, he said: 'Captain Smith, how are you?'

Then, according to Mr Pryal, the man answered: 'Very well, Pryal, but please don't detain me, I am on business.'

Hardly able to stand, so great was his astonishment, Mr Pryal, without realising what he was doing, followed the man to St Paul and Fayette Streets.

Several times the man turned, and when he finally saw Pryal behind him, rushed into the Calvert Building, and according to Mr Pryal, endeavoured to lose himself in the crowd. Pryal was behind him, however, and followed him through the Equitable Building and saw him board a west-bound car on Fayette Street.

His pursuer boarded the same car and saw the man get off at the Washington, Baltimore and Annapolis Station, where he purchased a ticket to Washington [DC]. As he passed through the gates to board he turned to Mr Pryal, smiled and said: 'Be good, shipmate, until we meet again.'

Mr Pryal when seen today said that he did not expect to be believed when he told of the incident, and added with great earnestness that he was willing to swear to his statements.

A third report of the incident published in *The Cincinnati Enquirer* on 21 July 1912 added:

The *Honolulu Star-Bulletin* for 3 August 1912 adds the details that, 'It was while on his way to the offices of Doctor Matier Warfield for treatment for an internal disorder last Wednesday that he swears he first saw approaching him the commander of the *Titanic*. Attired in a neat-fitting business suit of a light-brown colour, straw hat, and tan shoes, the man carried two suitcases and was staring straight ahead. Pryal approached him and spoke, but received no reply. The man seemed unconscious of his surroundings and continued walking rapidly west out of Baltimore Street.'

The Baltimore Enquirer, 21 January 1914 – 'Prayers Cure Cancer – Supplications to the Virgin Mother give Man Relief – he claims':

In answer to his prayers to the Virgin Mother for two years, a cancer on his nose, for which he has suffered for the last 27 years has been cured, declares Peter Pryal. Aged 72.

The old man said he retired one night, and on awakening he discovered that the cancer, which had been eating its way into his left eye and into his brain, had been cured. Pryal discarded the shield which he has worn over his nose for years and the skin of the nose was perfectly dry.

***

It, of course, has to be taken into account just how much an incipient cancer eating into the eye and brain influenced Captain Pryal's vision of the master of the *Titanic*? Would this have influenced recognition of his voice too?

Also, was it a coincidence that a man who was a long-time friend saw him two times after the disaster in a region of the United States associated with his step-family? Captain Smith's mother, Catherine – née Marsh (1808–1893), had married George John Hancock on 13 March

1831, but he died in about 1847. She then married Edward Smith (1804–1864), Captain Smith's father. In consequence of this, Smith had half siblings named Joseph, John and Thirza (or Thirra) Hancock. Joseph Hancock had at least two sons named George John (born 1858) and Frank Arthur Douglas (born 1862), and Captain Smith confirms in a letter of 1905 that Frank Hancock was his nephew, and was an obvious favourite of his. Frank Hancock was originally married to Annie Oaks Hatch, with whom he had eleven children, and after her death he remarried and had three children. He moved to the United States, where he lived in Savannah, New York, and had an office in the city. In the letter Captain Smith remarks, 'I gather Frank and family moved up to CT from Savannah.' The Hancock family were also known to have been associated with Baltimore and Washington DC, Captain Smith's destination on the train.

Were at least four men mistaken? Or could it be that Captain Smith was rescued in Fifth Officer Lowe's group of lifeboats, and being the captain of *Titanic* – 'The captain goes down with his ship' – which had taken a lot of their loved ones to the bottom of the ocean, was he taken away from the midst of the rescued passengers on *Carpathia* to avoid facing any awkward questions or even hostility.

If it was Captain Smith who was seen in Baltimore on his way to Washington DC, why did he not make it known that he had survived? The captain of a ship is the most responsible person on board the vessel and his primary role is to get all the passengers to their destination in safety. With this in mind, it could have been Bruce Ismay who ordered that he stayed quiet until they judged the reaction of the authorities and the general public. They were aware that there would be an inquiry, which would not be sympathetic towards Smith; probably leading to him being prosecuted if they thought he had survived. If he was alive he would have read the findings of the American inquiry in Washington DC, which was particularly scathing, to the extent that the Imperial Merchant Service Guild in Liverpool sent a letter to Senator Smith expressing their 'profound indignation' of his findings, and requesting that he retract much of his 'unwarrantable attack' on *Titanic*'s surviving officers. Indeed, Ismay came into severe criticism, and was even attacked. When the inquiry found that Smith was guilty of negligence, he probably decided not to speak out. He was aged 62 and had stated that he was ready for retirement, and as time went by it became more and more unethical to disclose his whereabouts – and a lot safer to stay silent.

Chapter 10

THE BRITISH *TITANIC* INQUIRY

The United States Senate held an inquiry into the sinking of *Titanic* in New York, from 19 April (only four days after the disaster) to 25 May 1912, chaired by Senator William Alden Smith. The inquiry was heavily criticised and ridiculed in Britain, so much so that the following information appeared in newspapers in September 1912:

> Merchants' Service Guild and the *Titanic* – Captain Mark Breach, New South Wales agent for the Imperial Merchant Service Guild of Liverpool, have received the following letter from the secretary:
>
> Now that the judgment of Lord Mersey in connection with the loss of the *Titanic* has been made public, the Imperial Merchant Service Guild have decided to publish a letter which was directed, by the Guild, to Senator Smith, some few weeks ago for his unwarrantable attack on the executive officers of the *Titanic*, in a speech made after the Committee's report had been introduced. Senator Smith not only criticised Captain Smith for failing to heed ice warnings, but he alleged that the junior officers availed themselves of the first opportunity to leave the ship. These and other baseless charges created considerable indignation throughout the whole of the mercantile world, particularly amongst captains and officers, and the Guild, in the interest of those concerned, deemed it necessary to address the following communication to Senator Smith:
>
> Imperial Merchant Service Guild, Liverpool, 8th July, 1912.
>
> Senator William Alden Smith.
>
> Sir, I am directed by the Guild, which is the great representative body of the captains and officers of the British Merchant Service, to inform you of the profound indignation, which prevails throughout that service, ranging from the highest to the lowest rank, at the malevolence which characterised your speech at Washington on the report of your committee,

which enquired into the loss of the *Titanic*. Had your recriminations and criticisms been confined to those really responsible for grave laxity in the way of obsolete regulations or the omission to make such in order to keep pace with the modern developments of merchant shipping, your committee and your remarks would have done a great public service, but when, without a shred of evidence or a particle of truth, you accuse officers of the British Mercantile Marine of despicable cowardice, we deem it requisite to repel such odious insinuations, whilst we would suggest that the falsity of them is such as to minimise very greatly the importance, which, otherwise, would have been attached to the report of your committee and your own speech upon it.

You speak of 'strong men used to command being rudely silenced by junior officers, who availed themselves of the first opportunity to leave the ship.'

These officers did nothing of the kind. It has been proved up to the hilt, that in the midst of an appalling emergency perfect discipline prevailed, which was worthy of the best traditions of the British Mercantile Marine; indeed, under such terrible circumstances, it is really a marvellous thing that so many were saved.

It is obvious to the merest tyro in shipping affairs that the [life]boats should, as far as possible, be in charge of responsible officers. The four officers who left the ship in charge of the four [life]boats did so, acting upon definite instructions from their superior officers, and, as you know, the three senior officers, and one junior officer gave up their lives like heroes, when, had they been possessed of a spark of cowardice, they had a better opportunity than any on board to get into the boats and save, at least, their own lives.

We sincerely trust that this matter having been directly brought to your notice by the Guild, we may either learn from you that a gross reflection on the honour of a great profession of this country has been made in error, or that, on the other hand, on further and more mature consideration, you will be good enough to withdraw it. In justice to yourself, we consider it to be the creature of impulse rather than of deliberation.

I am, sir, Your obedient servant,
[Signed] T.W. Moore, secretary.

The Guild on Friday last officially acquainted Senator Smith of Lord Mersey's version of the case:

Imperial Merchant Service Guild, Liverpool, 2nd August, 1912.

Senator William Alden Smith, Washington, U.S.A.

Sir, Supplementing the letter which the Guild addressed to you on the 8th July last, I am now directed to acquaint you of the following extract which appears in the judgment of Lord Mersey, in connection with the loss of the *Titanic*:

'The evidence satisfies me that the officers did their work very well, and without any thought of themselves. Captain Smith, the master, Mr Wilde, the chief officer, Mr Murdoch, and first officer, and Mr Moody, the sixth officer, all went down with the ship while performing their duties; the others, with the exception of Mr Lightoller, took charge of the boats, and thus were saved. Mr Lightoller was swept off the deck as the vessel went down, and was subsequently picked up.

'This, of course, fully justifies our letter to you, and our feelings of indignation at your suggestion of gross cowardice on the part of worthy members of the British Mercantile Marine.'

I am, sir, Your obedient servant,
[Signed] T.W. Moore, secretary.

The British Wreck Commissioner held an inquiry from 2 May until 3 July 1912, on behalf of the British Board of Trade, in the London Scottish Drill Hall at Buckingham Gate in Westminster, London. It was chaired by the High Court Judge Lord Mersey. The final report was published on 30 July 1912. Because the officers had time to think and remember more accurately, many of the statements varied from those that were given at the Senate inquiry.

Second Officer Charles Herbert Lightoller
Charles Lightoller was questioned in the United States on 19 April 1912, and re-called four days later on 23 April 1912. He was further and extensively questioned at the British inquiry on days 11, 12 (all day) and 14 as follows:

Day Eleven, 20 May 1912

Solicitor General: 'You are Mr Charles Herbert Lightoller, I think?'

Lightoller: 'Yes.'

Solicitor General: 'Were you second officer on *Titanic*?'

Lightoller: 'I was.'

Solicitor General: 'I think you hold a master's certificate?'

Lightoller: 'Yes.'

Solicitor General: 'You passed for master's in 1899?'

Lightoller: 'About that, yes.'

Solicitor General: 'And do you also hold an extra master's certificate?'

Lightoller: 'Yes.'

Solicitor General: 'Which you passed for in 1902?'

Lightoller: 'Yes.'

Solicitor General: 'How long have you been in the White Star Company's employ?'

Lightoller: 'Nearly 12½ years.'

Solicitor General: 'That would be about since 1900?'

Lightoller: 'January 1900.'

Solicitor General: 'Sailing with that company across the Atlantic many times, is most of your experience in the North Atlantic?'

Lightoller: 'Most, yes.'

Solicitor General: 'One other thing I should have asked you about your position; I think you do hold the position of first officer with the White Star?'

Lightoller: 'Yes.'

Solicitor General: 'But on this voyage you were second officer of the ship?'

Lightoller: 'Yes.'

Solicitor General: 'I will ask you the details later on, but I will ask you this now; were you present at the trial trip of *Titanic* in Belfast?'

Lightoller: 'Yes.'

Solicitor General: 'And I think, with the exception of Mr Wilde, all the officers whose names you have mentioned were present on that trial trip?'

Lightoller: 'Yes, they were.'

Solicitor General: 'And Mr Wilde joined you later?'

Lightoller: 'Yes.'

Solicitor General: 'Up to the time this vessel started her voyage from Southampton what was the greatest speed she had attained in practice?'

Lightoller: 'That is from Belfast round to Southampton we averaged about 18 knots.'

Solicitor General: 'That is the average. Do you know what was the greatest she had to go?'

Lightoller: 'Perhaps 18 and a half; I do not think she got much higher than that.'

Solicitor General: 'You left Southampton, as we know, on 10 April [1912], and you went across to Cherbourg?'

Lightoller: 'Yes.'

Solicitor General: 'And, I think, left Cherbourg about 9 o'clock on the 10th?'

Lightoller: 'About that.'

Solicitor General: 'And went to Queenstown?'

Lightoller: 'Yes.'

Solicitor General: 'When was it you left Queenstown?'

Lightoller: 'About 2 pm, as near as I can remember, on the following day.'

Solicitor General: 'On the 11th [April 1912]?'

Lightoller: 'Yes.'

Solicitor General: 'Just give me, if you will, the arrangement about the watches between the chief officer [Wilde], the first officer [Murdoch], and yourself. I suppose you would count as the three senior officers?'

Lightoller: 'Yes, exactly.'

Solicitor General: 'How was that?'

Lightoller: 'The chief officer [Wilde] had from 2 until 6 am and pm, the second officer [Lightoller] …'

Solicitor General: 'That is you?'

Lightoller: 'Yes, myself. The second officer [Lightoller] relieved the chief [Wilde] at 6 o'clock and was on deck until 10 – 6 to 10 am and pm. The first officer [Murdoch] was on deck from 10 to 2 am and pm.'

Solicitor General: 'Then the junior officers would be divided into watches, I suppose, and would serve with one or other of the seniors?'

Lightoller: 'They are divided into watches – 3 to 5 and 4 to 6, four hours on and four hours off, with a dog watch, that is, the watch from 4 to 8 pm, is divided into what we call the dog watches, 4 to 6 and 6 to 8.'

Solicitor General: 'We will go to Sunday, 14 April [1912]. Your first watch, the morning watch, would be from 6 to 8 am, as I follow you?'

Lightoller: 'Yes.'

Solicitor General: 'Then, having completed that watch, do you come to the bridge again after luncheon time?'

Lightoller: 'Yes.'

Solicitor General: 'Just tell us about it?'

Lightoller: 'Lunch is half-past twelve. I relieve the first officer [Murdoch], who has his lunch at half-past twelve, and he comes on deck again about one o'clock or five minutes past; then I have mine.'

Solicitor General: 'It really means that there is half-an-hour out of the first officer's [Murdoch's] watch? Now, on this day, 14 April [1912], did you follow that course?'

Lightoller: 'Yes.'

Solicitor General: 'And relieved Mr Murdoch from 12.30 pm to about 1 pm?'

Lightoller: 'Yes.'

Solicitor General: 'Do you remember Captain Smith showing you something during that time?'

Lightoller: 'Yes.'

Solicitor General: 'Just tell us what it was?'

Lightoller: 'Captain Smith came on the bridge during the time I was relieving Mr Murdoch. In his hands he had a wireless message, a Marconigram. He came across the bridge, and holding it in his hands he told me to read it.'

Solicitor General: 'He showed it to you, I suppose?'

Lightoller: 'Yes, exactly; he held it out in his hand and showed it to me. The actual wording of the message I do not remember.'

Solicitor General: 'Did you see whether it was about ice?'

Lightoller: 'It had reference to ice.'

Solicitor General: 'Do you remember between what meridians?'

Lightoller: 'I particularly made a mental note of the meridians – 49 to 51.'

Solicitor General: 'That would be 49 to 51 west?'

Lightoller: 'Yes.'

Solicitor General: 'Exactly.'

Sir Robert Finlay: 'We have the message. I will just find it and read it to you, and perhaps you will be able to tell me if that is right. Do you know from what ship the message came?'

Lightoller: 'I cannot remember the ship.'

Solicitor General: 'It is better to have it now.'

Sir Robert: 'Yes, I think we had better have it, and the ship it came from.'

Solicitor General: 'My recollection is that the attorney general read it in opening.'

Commissioner: 'What time was it?'

Solicitor General: 'So far, My Lord, he has said it was between 12.30 and one o'clock in the middle of the day.'

Solicitor General: [To Lightoller] 'Can you fix at all as between those times?'

Lightoller: 'About 12.45 as near as I can remember.'

Solicitor General: 'Very well, about a quarter to one?'

Lightoller: 'Yes.'

F. Laing: 'I have the wording of it.'

Solicitor General: 'I think this is the message, and perhaps I can read it to the gentleman and he will tell us if it sounds like it.'

Solicitor General: [To Lightoller] 'We have independent evidence of a message being sent from the [RMS] *Caronia*: "West-bound steamers report [ice]bergs, growlers and field ice in 42 north; from 49 to 51 west"?'

Lightoller: 'I think that is the message that I referred to as near as I can remember.'

Solicitor General: 'The witness says he was shown that at about quarter to one, My Lord, your Lordship will find the evidence of Captain Barr, the captain of the *Caronia*, who was interposed on Friday ... The attorney general asked Captain Barr, "On the morning of 14 April [1912], that is, on the Sunday morning, do you remember sending this Marconigram to *Titanic*?" Westbound steamers report bergs, growlers and field ice in 42 north, from 49 to 51 west?'

Captain Barr: "Yes, I remember sending it."

Attorney General: "That is sent, I see from your notes, at nine o'clock in the morning."

Captain Barr: "That is the time when the message was sent from *Caronia*."'

Commissioner: 'Does it go on to say that the message was acknowledged?'

Solicitor General: 'Yes, My Lord.'

'Attorney General: "And did you receive a reply at 9.44 am?"

Captain Barr: "Yes, as per that statement. The reply from Captain Smith was: 'Thanks for message and information – have had variable weather throughout – Smith'."'

Solicitor General: [To Lightoller] 'Now, the *Caronia* as we know was coming from New York to Europe, and as you see there is the message. The acknowledgement is 9.44 am *Caronia's* time. You had not heard anything about that before you went off your watch at 10 o'clock?'

Lightoller: 'No.'

Solicitor General: 'Can you help us? Would 9.44 am *Caronia*'s time coming from New York be likely to be later than your 10 o'clock watch coming to an end? You say you went off duty at ten.'

Lightoller: 'Yes.'

Commissioner: 'Did Captain Smith tell you when he had received the Marconigram?'

Lightoller: 'No, My Lord.'

Solicitor General: 'And the first you knew of it was when Captain Smith showed it to you at about quarter to one?'

Lightoller: 'Yes.'

Solicitor General: 'So far as your knowledge goes is that the first information as to ice which you had heard of as being received by *Titanic*?'

Lightoller: 'That is the first I have any recollection of.'

Solicitor General: 'What time of day would it be that your ship's course would be set?'

Lightoller: 'At noon.'

Solicitor General: 'Would that be done by the commander [Smith]?'

Lightoller: [Did not answer.]

Solicitor General: 'Add anything if there is anything we ought to know. Is that the incident as it occurred then?'

Lightoller: 'That is the whole of the incident, when the commander [Smith] came out and showed me the wireless, yes.'

Solicitor General: 'And you told us you were relieving Mr Murdoch while he was away at lunch. Did he come back?'

Lightoller: 'Yes. When he came back I mentioned the ice to him.'

Solicitor General: 'When you mentioned this message about the ice to Mr Murdoch when he came back at one o'clock did you gather from Mr Murdoch that it was news to him or did you gather from him that you had heard it before?'

Lightoller: 'That I really could not say, whether it was fresh news to him or not; I should judge that it would have been, but I really could not say from his expression – not from what I remember.'

Solicitor General: 'Your impression is that it was news to him?'

Lightoller: 'Probably.'

Solicitor General: 'Then did you leave the bridge at that time?'

Lightoller: 'Yes.'

Solicitor General: 'And your watch of course would not return until six in the evening?'

Lightoller: 'Exactly.'

Commissioner: 'Can you tell me what the ship's course was at that time?'

Lightoller: 'The compass course?'

Commissioner: 'Yes.'

Lightoller: 'No, I cannot remember what it was.'

Solicitor General: 'You are able to tell us late in the day what it was?'

Lightoller: 'The true course.'

Solicitor General: 'Can you tell us the true course of the ship at this time?'

Lightoller: 'No, I am afraid I cannot.'

Solicitor General: 'Here was a message shown you which referred to ice in latitude 42 N [north]?'

Lightoller: 'Yes.'

Solicitor General: 'Do you recollect, or can you help us at all, did that indication 42 N [north] indicate to you that it was near where you were likely to go?'

Lightoller: 'It would, had I taken particular notice of the latitude, though as a matter of fact, latitude as regard to ice conveys so very little.'

Solicitor General: 'Is that because it tends to set north or south?'

Lightoller: 'North and south, yes.'

Commissioner: 'I do not follow that?'

Lightoller: 'We take very little notice of the latitude because it conveys very little. You cannot rely on latitude.'

Solicitor General: 'For ice?'

Lightoller: 'Yes.'

Solicitor General: 'He answered that "because the ice tends to set north and south."'

Solicitor General: [To Lightoller] 'Then do you attach more importance to the longitude?'

Lightoller: 'Far more.'

Solicitor General: 'I notice your recollection of the message is you recollect 49 and 51 west?'

Lightoller: 'Distinctly.'

Solicitor General: 'That is longitude. Did you form any sort of impression at that time as to what time of day or night you were likely to reach the area indicated?'

Lightoller: 'Not at that time.'

Solicitor General: 'I know you worked it out, or helped to work it out later?'

Lightoller: 'It was worked out.'

Solicitor General: 'But you did not form any opinion at the time?'

Lightoller: 'Not at the time.'

Solicitor General: 'As far as you are concerned is there anything you deem important to tell us as between one o'clock and six o'clock when you came on duty?'

Lightoller: 'No, I cannot remember anything of importance.'

Commissioner: 'At the time this message was given to you by Captain Smith, how many hours steaming would you be away from the ice-field?'

Lightoller: 'I did not calculate the time; later I told one of the junior officers to work out about what time we should reach the ice region, and he told me about eleven o'clock.'

Commissioner: 'At night?'

Lightoller: 'This was after I came on deck again thought, at 6 o'clock, I knew that we should not be in the vicinity of the ice before I came on deck again. I roughly ran that off in my mind.'

Solicitor General: 'That is what I meant?'

Lightoller: 'Yes, I ran that roughly off in my mind.'

Solicitor General: 'When you saw this message at a quarter to one you saw that it was important but thought the position could not be reached until your watch came round again?'

Lightoller: 'I was sure of that.'

Solicitor General: 'You came on duty again at six o'clock?'

Lightoller: 'At 6 o'clock.'

Solicitor General: 'In the afternoon. That would be to relieve Chief Officer Wilde, as I follow you?'

Lightoller: 'Yes.'

Solicitor General: 'Did he hand the ship over to you at 6 o'clock?'

Lightoller: 'At 6 o'clock, yes.'

Solicitor General: 'Can you tell us what was the course of the ship when she was handed over to you at 6 o'clock?'

Lightoller: 'I cannot remember the compass course. I know from calculations made afterwards that we were making S [south] 86 true.'

Solicitor General: 'S [south] 86 W [west]?'

Lightoller: 'Yes.'

Solicitor General: 'That is within four degrees of due west true?'

Lightoller: 'Yes.'

Commissioner: 'Give me that again.'

Solicitor General: 'S [south] 86 W [west]. That is only four degrees from due west.'

Commissioner: 'Is it what I should call making a westerly course?'

Solicitor General: 'My Lord, I think I am right, and Sir Robert confirms me. The quartermaster at the wheel who gave evidence, who was at the wheel at the time of the disaster, said he was steering by compass a course of N [north] 71 W [west], so presumably N [north] 71 W [west]. Is the same thing as what this gentleman speaks of as S [south] 86 W [west], true.'

Lightoller: 'Pretty nearly. The compass course is not the compass we go by. I believe by standard we were steering N [north] 73. 86 true. I know it was, and I think that works out as 73 by compass, and 71 was the steering compass.'

Solicitor General: 'Did you learn whether while you had been off duty during the afternoon any further information had reached *Titanic* about ice?'

Lightoller: 'Not that I remember.'

Solicitor General: 'Of course, in the ordinary course, Mr Wilde would pass on to you any information that was necessary to help you during your watch?'

Lightoller: 'Yes.'

Solicitor General: 'And you have told us what happened?'

Lightoller: 'Yes.'

Solicitor General: 'Now what did you notice about the speed of your vessel?'

Lightoller: 'As far as I could tell her speed was normal.'

Solicitor General: 'Were they telegraphed at full speed ahead?'

Lightoller: 'At full speed.'

Commissioner: 'What do you mean by normal?'

Lightoller: 'Full speed.'

Commissioner: 'What is full speed; can you give me how many knots?'

Lightoller: 'We were steaming, as near as I can tell from what I remember of the revolutions – I believe they were 75 – and I think that works out at about 21 and a half knots the ship was steaming.'

Solicitor General: 'Is it the regular course for a message to be sent to the engine room from time to time, and a report to be got as to how many revolutions she is making?'

Lightoller: 'As a rule, at the end of the watch, the junior officer rings up the engine room and obtains the average revolutions for the preceding watch.'

Solicitor General: 'And is that one of the matters that would be brought before your notice when you go on duty?'

Lightoller: 'No, not necessarily. It is entered up in the logbook, and anyone who wishes to know can merely ask and the information is given him.'

Solicitor General: 'When you say your recollection is that it was 75 revolutions, just help us. What is it you have in your mind?'

Lightoller: 'I could not say where I got that from, but it is in my mind that it was about 75 revolutions.'

Solicitor General: 'In the course of the voyage across the Atlantic, had the engines, as far as you know, exceeded 75 at any time?'

Lightoller: 'On one occasion I have a recollection of one side turning 76, not necessarily both sides though.'

Solicitor General: 'That would be one or other of the sets of reciprocating engines?'

Lightoller: 'Port or starboard reciprocating, yes.'

Solicitor General: 'Subject to that as far as you know, did she ever attain a greater number of revolutions than 75?'

Lightoller: 'Not to my knowledge, and I think I should have heard of it if she had.'

Solicitor General: 'And during your watch which extended from 6 till 10, did she maintain the same speed, as far as you know?'

Lightoller: 'As far as I know.'

Solicitor General: 'Then who would be on the bridge – is it one or two of the junior officers would be on the bridge with you?'

Lightoller: 'Two junior officers on watch at all times.'

Solicitor General: 'There would be a quartermaster at the wheel?'

Lightoller: 'And a stand-by quartermaster.'

Solicitor General: 'Another quartermaster standing by?'

Lightoller: 'Exactly.'

Solicitor General: 'And there would be two look-out men in the crow's nest?'

Lightoller: 'At all times.'

Solicitor General: 'What was the practice in the "*Titanic*" as far as this voyage is concerned about having a look-out man anywhere else?'

Lightoller: 'In anything but clear weather we carry extra look-outs.'

Solicitor General: 'But where do you put them?'

Lightoller: 'If the weather is fine, that is to say if the sea allows it, we place them near the stem head; when the weather does not allow us placing them at the stem head, then probably on the bridge.'

Solicitor General: 'And as far as your watch was concerned, 6 to 10 on the evening of April 14th [1912], was there any look-out except the two men in the crow's-nest?'

Lightoller: 'No.'

Solicitor General: 'What was the weather?'

Lightoller: 'Perfectly clear and fine.'

Solicitor General: 'Had there been, as far as you remember, any occasion since she left Southampton to have extra look-out men?'

Lightoller: 'Yes, and we had had them.'

Solicitor General: 'And you had had them?'

Lightoller: 'Yes.'

Solicitor General: 'But at this time it was clear and fine?'

Lightoller: 'Yes.'

Solicitor General: 'Of course the sea was calm?'

Lightoller: 'Comparatively smooth.'

Solicitor General: 'Could you see the stars?'

Lightoller: 'Perfectly clear. There was not a cloud in the sky.'

Solicitor General: 'There was no moon, I think?'

Lightoller: 'No moon.'

Solicitor General: 'During your watch was any change made in the course?'

Lightoller: 'Not to my recollection.'

Solicitor General: 'Then when you had taken the ship over from Mr Wilde and gathered this information, I think you gave some directions to one of the junior officers?'

Lightoller: 'I directed the sixth officer to let me know at what time we should reach the vicinity of the ice. The junior officer reported to me, "About 11 o'clock."'

Solicitor General: 'Do you recollect which of the junior officers it was?'

Lightoller: 'Yes, Mr Moody, the sixth.'

Solicitor General: 'That would involve his making some calculations, of course?'

Lightoller: 'Yes.'

Solicitor General: 'Had this Marconigram about the ice with the meridians on it been put up; was it on any notice board, or anything of the sort?'

Lightoller: 'That I could not say with any degree of certainty. Most probably, in fact very probably, almost certainly, it would be placed on the notice board for that purpose in the chart room.'

Solicitor General: 'At any rate when you gave Mr Moody those directions he had the material to work on?'

Lightoller: 'Exactly.'

Solicitor General: 'And he calculated and told you about 11 o'clock, you would be near the ice?'

Lightoller: 'Yes.'

Solicitor General: 'That is to say an hour after your watch finished?'

Lightoller: 'Yes. I might say as a matter of fact I have come to the conclusion that Mr Moody did not take the same Marconigram which Captain Smith had shown me on the bridge because on running it up just mentally, I came to the conclusion that we should be to the ice before 11 o'clock, by the Marconigram that I saw.'

Commissioner: 'In your opinion when in point of fact would you have reached the vicinity of the ice?'

Lightoller: 'I roughly figured out about half past nine.'

Commissioner: 'Then had Moody made a mistake?'

Lightoller: 'I should not say a mistake, only he probably had not noticed the 49 degree wireless; there may have been others, and he may have made his calculations from one of the other Marconigrams.'

Commissioner: 'Do you know which other Marconigrams he would have to work from?'

Lightoller: 'No, My Lord. I have no distinct recollection of any other Marconigrams.'

Commissioner: 'Because it is suggested to me that there was no Marconigram which would indicate arrival at the ice-field at 11 o'clock?'

Lightoller: 'Well, My Lord, as far as my recollection carries me, Mr Moody told me 11, and I came to that conclusion that he had probably used some other Marconigram.'

Commissioner: 'It did not agree with your conclusion?'

Lightoller: 'No.'

Solicitor General: 'Your Lordship will find in the print, at pages 12 and 13, when the attorney general was opening another Marconigram from the Baltic. I would like to follow this a little. I think My Lord will agree.'

Solicitor General: [To Lightoller] 'You have just said you came to the conclusion that Mr Moody had been working on some message other than the one Captain Smith had shown you?'

Lightoller: 'Exactly.'

Solicitor General: 'When he came to you on your watch – of course, you are responsible up to 10 o'clock?'

Lightoller: 'Yes.'

Solicitor General: 'When he came to you on your watch and said you would get to the ice, as he calculated about 11, did you, as far as you remember, say anything to him about it?'

Lightoller: 'No.'

Solicitor General: 'It was important to you?'

Lightoller: 'I quite see your point, and I had reasons for not doing so. As far as I remember he was busy – what on I cannot recollect, and I thought I would not bother him just at that time. He was busy with some calculations, probably stellar calculations or bearings, and I had run it up in my mind, and I was quite assured that we should be up to 49 degrees somewhere about half past 9.'

Day Twelve, 21 May 1912 (all day)

Solicitor General: 'You were just telling us what you found when you came up on deck after you had heard of what had happened, and I think you just told us that the steam was roaring off – blowing out of the boilers, I suppose?'

Lightoller: 'Yes.'

Solicitor General: 'Was it making a great noise?'

Lightoller: 'Yes.'

Solicitor General: 'So great as to be difficult to hear what was said?'

Lightoller: 'Very difficult.'

Solicitor General: 'Did you ascertain whether all hands had been called on deck?'

Lightoller: 'Yes; I met the chief officer [Wilde] almost immediately after, coming out of the door of the quarters. First of all the chief officer told me to commence to get the covers off the [life]boats. I asked him then if all hands had been called, and he said, "Yes."'

Solicitor General: 'I should like to understand whether there was a division of duties here. In an emergency of this sort, have you a special responsibility for one side of the ship as against the other?'

Lightoller: 'No.'

Solicitor General: 'Then there is an order from the chief officer that you should see to the stripping of the covers off the [life]boats?'

Lightoller: 'Yes.'

Solicitor General: 'Did you do that?'

Lightoller: 'Yes.'

Solicitor General: 'At that time had any of the [life]boats had their covers stripped, or had you to begin it?'

Lightoller: 'None, with the exception of the emergency [life]boats.'

Solicitor General: 'Those were the two which we have heard of which were kept swung out?'

Lightoller: 'Yes.'

Solicitor General: 'And did you get hands to help you in that work?'

Lightoller: 'Yes, I commenced myself, and then as the hands turned up, I told them off to the [life]boats.'

Solicitor General: 'Which side did you begin, and what was the order?'

Lightoller: 'I began on the port side with the port forward [life]boat. That would be number 4.'

Solicitor General: 'That would be the one immediately abaft of the emergency [life]boat?'

Lightoller: 'Yes.'

Solicitor General: 'Just tell us the order of things, will you?'

Lightoller: 'I commenced stripping off [lifeboat] number 4; then two or three [men] turned up; I told them off to number 4 [life]boat and stood off then myself and directed the men as they came up on deck, passing around the boat deck, round the various [life]boats, and seeing that the men were evenly distributed around both the port and starboard.'

Solicitor General: 'Do you mean evenly distributed as between the different [life]boats?'

Lightoller: 'Exactly.'

Solicitor General: 'Had you any means of knowing what [life]boat a particular seaman would be attached to if he did not know; have you any means of telling him?'

Lightoller: 'Well, I did not think it advisable, taking into consideration the row going on with the steam to make any inquiries. I could only direct them by motions of the hand. They could not hear what I said.'

Solicitor General: 'So that you parcelled them out as best you could?'

Lightoller: Exactly.'

Solicitor General: 'Did you go to [life]boats in the afterend as well?'

Lightoller: 'Yes.'

Solicitor General: 'On the port side?'

Lightoller: 'Both sides.'

Solicitor General: 'Then you went the whole circuit of the boat deck?'

Lightoller: 'Yes.'

Solicitor General: 'Carrying out this order?'

Lightoller: 'Yes.'

Solicitor General: 'And was each of the [life]boat covers stripped in order all the way round.'

Lightoller: 'All the [life]boats, as far as I can remember, were under way. I remember directing one of the junior officers to look after the after section of [life]boats.'

Solicitor General: 'What length of time would this operation of uncovering all these boats take?'

Lightoller: 'You mean, given the crew?'

Solicitor General: 'You were engaged on this work. I want to realise how long you were engaged on it?'

Lightoller: 'Well, I really could not say what time the after [life]boats were finished uncovering. Knowing that the third officer [Pitman] was there in charge I did not bother so much about that as the forward ones, and about the time I had finished seeing the men distributed round the deck, and the [life]boat covers well under way and everything going smoothly, I then enquired of the chief officer [Wilde] whether we should carry on and swing out.'

Solicitor General: 'And what did Mr Wilde say about that – what were the orders?'

Lightoller: 'I am under the impression that Mr Wilde said "No", or "Wait," something to that effect, and meeting the commander [Smith], I asked him, and he said, "Yes, swing out.".'

Solicitor General: 'And did you get that done?'

Lightoller: 'Yes, on the port side. I did not go to the starboard side again.'

Solicitor General: 'Up to the time of swinging out the [life]boats which had been stripped, at any rate, on the port side, what about the passengers?'

Lightoller: 'I had met a few passengers on deck, not many.'

Solicitor General: 'Had you heard any general orders given about getting them?'

Lightoller: 'No, I could not hear any.'

Solicitor General: 'Was the steam still blowing off all this time?'

Lightoller: 'Still blowing off, yes.'

Solicitor General: 'Up to this time had you noticed whether the ship had got any list?'

Lightoller: 'Not to my knowledge; no list whatever so far as I know.'

Solicitor General: 'Up to this time had you noticed whether she showed a tendency to drop by the head?'

Lightoller: 'No.'

Solicitor General: 'She was on an even keel so far as you know?'

Lightoller: 'Yes.'

Commissioner: 'Now, you say "at this time". I do not quite know what time.'

Solicitor General: 'I was stopping; I had meant to stop at the time he ceased to swing out [life]boats on the port side, which is, as I understand, after stripping all the [life]boat covers.'

Commissioner: 'I understand about the course of events at this time; I want to know by the clock.'

Solicitor General: 'I did, too, My Lord.'

Solicitor General: [To Lightoller] 'Could you help us and give us some estimate as to how long this would have taken from the time that you came out. You see, you have said you think half an hour elapsed after the collision before you came out and realised the seriousness of it, and then, of course, you undertook those duties, and you have described them. Could you give us an estimate how long would have elapsed from the time you came out on deck and started this work to the time the [life]boats were swung out on the port side?'

Lightoller: 'I should like you to understand quite clearly about the [life]boat covers. I had not seen all the [life]boat covers actually off. We were taking the [life]boats in rotation, but from the time we commenced to strip number 4 [life]boat cover until the time when we swung them out I should judge would be probably at most 15 or 20 minutes.'

Solicitor General: 'So far you are confining yourself to [lifeboat] number 4?'

Lightoller: 'Exactly.'

Solicitor General: 'And during that time had the stripping of the covers of the other [life]boats been going on?'

Lightoller: 'That was being continued at the same time. Of course, there were the falls to coil down.'

Solicitor General: 'You took [lifeboat] number 4. Was the swinging out of number 4 earlier than the swinging out of the other [life]boats on the port side?'

Lightoller: 'Yes, as it happened. You see the men coming up the staircase on the forepart would naturally come to [lifeboat] number 4, and number 4 was got under way first and would be completed first.'

Solicitor General: 'Did you go on your way down the port side getting it done?'

Lightoller: 'Yes.'

Solicitor General: 'Taking the swinging out of the last [life]boat that you saw to on the port side, how much later would that be?'

Lightoller: 'That was very late on.'

Commissioner: 'That is what I want to know?'

Lightoller: 'Well, you see, if I may give it to you in the order that I was working, I swung out [lifeboat] number 4 with the intention of loading all the [life]boats from A Deck, the next deck below the boat deck. I lowered [lifeboat] number 4 down to A Deck, and gave orders for the women and children to go down to A Deck to be loaded through the windows. My reason for loading the [life]boats through the windows from A Deck was that there was a coaling wire, a very strong wire running along A Deck, and I thought it would be very useful to trice the [life]boat to in case the ship got a slight list or anything; but as I was going down the ladder after giving the order, someone sung out and said the windows were up. I countermanded the order and told the people to come back on the boat deck and instructed two or three, I think they were stewards, to find the handles and lower the windows. That left number 4 [life]boat hanging at A Deck, so then I went on to [lifeboat] number 6.'

Solicitor General: 'And was [lifeboat] number 6 still on the boat deck?'

Lightoller: 'Yes. Then I proceeded to put out [lifeboat] number 6 and lower away. Previous to this, I may say I had had orders from the commander [Smith] to fill the [life]boats with women and children, put women and children into the [life]boats and lower away.'

Solicitor General: 'Of course, the model we have there shows the starboard side, but the arrangement is the same for this purpose, I think, and one sees that if one took the [life]boat immediately abaft the emergency [life]boat, and lowered it to A Deck, it would in that model come against the closed-in side?'

Lightoller: 'Yes.'

Solicitor General: 'With the windows in it?'

Lightoller: 'Yes.'

Solicitor General: 'And your idea was that those windows should be opened and the people should get from the windows into the [life] boat from A Deck?'

Lightoller: 'Exactly.'

Solicitor General: 'Then that plan was not, in fact, carried out?'

Lightoller: 'No, not on the port side.'

Solicitor General: 'For the reason you have explained?'

Lightoller: 'Yes.'

Commissioner: 'I am still without the information I want.'

Solicitor General: 'I realise that, and I will come back to it.'

Commissioner: [To Lightoller] '[Lifeboat] number 4 would take some time. Then what was the next [life]boat, so far as you are concerned, which was filled with women and children?'

Lightoller: '[Lifeboat] number 6.'

Solicitor General: 'And the next one?'

Lightoller: 'As far as I remember, [lifeboat] number 8.'

Solicitor General: 'That exhausts the [life]boats, which are forward, on the port side?'

Lightoller: 'Yes.'

Solicitor General: 'Then did you see to the loading of any others [lifeboats] on the port side?'

Lightoller: 'I went forward – the last lifeboat for me to load on the port side was [lifeboat] number 4 from A Deck.'

Solicitor General: 'It got as far as that?'

Lightoller: 'Yes, and it remained there.'

Solicitor General: 'Now what I want to know is this – making the best estimate you can, can you give us some help as regards the time – either the time which had elapsed, or the time by the clock when the lowering away of [lifeboat] number 4 actually took place, putting it into the water?'

Lightoller: 'Would it be of any assistance, if I gave you the time that the collapsible [life]boat, the actual last [life]boat, got away on the port side?'

Commissioner: 'Well, it might.'

Lightoller: 'I can remember that distinctly – lowering it only about 10 feet.'

Commissioner: 'I will tell you what I want, and then perhaps you will be able to answer. You said that after the [life]boats on the port side had been lowered the ship had no list, either to port or starboard, and that she was not down by the head. Now, I want to know at what time you observed that?'

Solicitor General: 'What I understand him to say was that the [life] boats were swung out before he had noticed it. I did not understand him to say that they were lowered.'

Albion House near Pier Head in Liverpool was the headquarters of the White Star Line from 1898 until 1934. It was from the balconies on the second floor, partly obscured by the iron bridge, that officials made announcements concerning the *Titanic* disaster, too afraid to leave the building fearing the reaction of the devastated crowds.

RMS *Oceanic II* was launched in 1899. Officers Charles Lightoller, Herbert Pitman and James Moody all served on her prior to joining *Titanic*.

RMS *Titanic* leaves Belfast to travel to Southampton under the command of Captain Herbert 'Bert' James Haddock, with six of the seven officers who would be involved in the disaster on board.

RMS *Titanic* leaves Cherbourg on its way to Queenstown (now Cobh) on the south coast of Ireland.

Captain Edward John Smith from Stoke-on-Trent was aged 61, and had served with the White Star Line for over thirty years when he joined *Titanic*, with a not too impressive safety record. He had good reason to disappear into obscurity, and could it really have been a case of misidentification of both sight and voice?

Chief Officer Lieutenant Henry 'Harry' Tingle Wilde (1872–1912).

First Officer Lieutenant William 'Bill' McMaster Murdoch (1873–1912).

Second Officer Sub-Lieutenant Charles 'Charlie' Herbert Lightoller (1874–1952).

Third Officer Mr Herbert 'Bert' John Pitman MBE (1877–1961).

Fourth Officer Sub-Lieutenant Joseph 'Joe' Groves Boxhall (1884–1967).

Fifth Officer Sub-Lieutenant Harold Godfrey Lowe (1882–1944).

Sixth Officer Sub-Lieutenant James 'Jim' Paul Moody (1887–1912).

Many passengers were in two minds whether to get into the lifeboats. To them the choice was to remain on a massive ship, which was supposed to be 'unsinkable', or sail out into the cold, dark night on a small uncomfortable lifeboat, with no heat or light, and even though the sea was calm, it could still get into difficulties.

Lifeboat number 14 under the command of Fifth Officer Lowe as it approached *Carpathia*, with collapsible lifeboat D in tow. The picture gives a good impression of how isolated they must have felt in the vastness of the sea.

Proud of the courageous local William Murdoch, the people of Dalbeattie had a memorial dedicated to his name placed in the town hall, and they requested an apology from the makers of the 1997 James Cameron film *Titanic*, voicing fierce objections to the way he was incorrectly portrayed.

The memorial statue to Captain Smith was created in 1914 by Kathleen Scott, wife of Captain Robert Falcon Scott of the Antarctic. It has been debated for over 100 years why it was erected at Beacon Park in Lichfield, and not nearer to his birthplace at Hanley in Stoke-on-Trent.

Commissioner: 'I understand him to say that it was quite a long time.'

Solicitor General: 'Quite.'

Commissioner: 'I do not care whether they were lowered. At what time was it you noticed this ship had no list, and that it was not down by the head?'

Lightoller: 'When I came on deck and commenced uncovering the [life]boats.'

Commissioner: 'I understood you were speaking of a much later period.'

Solicitor General: [To Lightoller] 'I was asking about a later period?'

Lightoller: 'I am sorry.'

Solicitor General: 'When you came out on deck, having been aroused, the ship was on an even keel?'

Lightoller: 'Yes.'

Solicitor General: 'You had heard that the water was out up to F Deck?'

Lightoller: 'Yes.'

Solicitor General: 'But you did not notice any list?'

Lightoller: 'No.'

Solicitor General: 'How long did that state of things continue? When was it you did first notice either a list or that she was down by the head?'

Lightoller: 'Very shortly, afterwards I noticed she was down by the head, when I was by number 6 [life]boat. When I left [lifeboat] number 4 and went to number 6 she was distinctly down by the head, and I think it was while working at that [life]boat it was noticed that she had a pretty heavy list to port.'

Commissioner: 'This must have been within a quarter of an hour from your coming on the boat deck?'

Lightoller: 'No, My Lord, it would take us a quarter of an hour or 20 minutes to get [lifeboat] number 4 uncovered and the falls out.'

Commissioner: 'But when you did get [lifeboat] number 4 out you noticed this list, I understand?'

Lightoller: 'No, My Lord, I think I said at [lifeboat] number 6.'

Commissioner: 'Then how long would it take you to get [lifeboat] number 4 and number 6 uncovered?'

Lightoller: 'Well, it would take us from 15 minutes to 20 minutes to uncover [lifeboat] number 4; then to coil the falls down, then to swing out and lower it down to A Deck would take another six or seven minutes at least. Then I gave an order to go down to the lower deck which I countermanded; perhaps two or three minutes might have elapsed there. Then I went to [lifeboat] number 6 about that time.'

Commissioner: 'How long were you working at [lifeboat] number 6?

Lightoller: I really could not say, My Lord. I went to [lifeboat] number 6 then, as far as I remember.'

Commissioner: 'At what point of these events did you notice that the ship had begun to be down by the head or to have a list?'

Lightoller: 'It was when I was at number 6 [life]boat, My Lord.'

Commissioner: 'As I understand, that would be about half-an-hour after you had come on deck?'

Lightoller: 'I think it is longer than that.'

Commissioner: 'Well, let us say three quarters of an hour?'

Lightoller: 'Yes, perhaps three quarters of an hour.'

Commissioner: 'You had been half an hour in your bunk before you came on deck at all?'

Lightoller: 'I said approximately half an hour.'

Commissioner: 'So this would be an hour or an hour and a quarter after the collision. And was it then for the first time you noticed the vessel had a list?'

Lightoller: 'At whatever time that was, My Lord. However, it works out it was about when I was at [life]boat number 6.'

Solicitor General: 'What you had been doing in the interval was, you had been getting [lifeboat] number 4 unstripped; you had been getting her swung out, her falls cleared and let down as far as the A Deck, and there you had ascertained that it was not possible to open the windows and get the people through?'

Lightoller: 'Not immediately, and therefore rather than delay I did not go on with it.'

Solicitor General: 'That is what happened?'

Lightoller: 'Yes.'

Solicitor General: 'Then turning your attention to [lifeboat] number 6 you then noticed the ship had got a list?'

Lightoller: 'Yes, I think it was [lifeboat] number 6.'

Commissioner: 'And it was a list to port?'

Lightoller: 'Yes.'

Commissioner: 'Did you ever notice a list to starboard?'

Lightoller: 'No.'

Commissioner: 'Was there a list to starboard?'

Lightoller: 'Not that I am aware of, and I think I should have noticed it in lowering the [life]boat. I may say that my notice was called to this list – I perhaps might not have noticed it; it was not very great – by Mr Wilde calling out "All passengers over to the starboard side." That was an endeavour to give her a righting movement, and it was then I noticed that the ship had a list. It would have been far more noticeable on the starboard side than on the port.'

Solicitor General: 'Did you hear that order given when you were dealing with [life]boat number 6?'

Lightoller: 'Yes.'

Solicitor General: 'Now by that time you were dealing with [life]boat number 6, were there a number of passengers, men, women and children, on the [life]boat deck?'

Lightoller: 'Yes.'

Solicitor General: 'And at that time when you were dealing with [lifeboat] number 6 had any order been given about their getting into the [life]boats?'

Lightoller: 'Yes.'

Solicitor General: 'Who gave it, and when was it given?'

Lightoller: 'The captain gave it to me.'

Solicitor General: 'What was the order?'

Lightoller: 'After I had swung out number 4 [life]boat I asked the chief officer [Wilde] should we put the women and children in, and he said "No." I left the men to go ahead with their work and found the commander [Smith], or I met him and I asked him should we put the women and children in, and the commander said "Yes, put the women and children in and lower away." That was the last order I received on the ship.'

Solicitor General: 'Was that, as you understood it, a general order for the [life]boats?'

Lightoller: 'Yes, a general order.'

Solicitor General: 'Again, I should like to have the time fixed. Is that after these events you have described about [life]boat number 4?'

Lightoller: 'No; previous to any swinging out, when [lifeboat] number 4 was almost uncovered; in fact, the canvas cover was off. They were taking the falls out and I think they were in the act of taking the strong back out, and the next movement to be executed would be swinging the [life]boat out. So before any delay had occurred I asked the commander [Smith], as I say, should we lower away.'

Solicitor General: 'That means, should you put people into the boat, I suppose?'

Lightoller: 'Yes. We had had orders to swing out, so the [life]boat was in the process of being swung out.'

Solicitor General: 'Now, we can take [lifeboat] number 6. You say you went to that?'

Lightoller: 'Yes.'

Solicitor General: 'You saw that [life]boat filled, did you?'

Lightoller: 'Yes.'

Solicitor General: 'It was filled under your supervision?'

Lightoller: 'Yes.'

Solicitor General: 'Now, tell us about the way in which it was done and the orders given as to who should get into it?'

Lightoller: 'As a matter of fact, I put them in myself. There were no orders. I stood with one foot on the seat just inside the gunwale of the [life]boat, and the other foot on the ship's deck, and the women merely held out their wrist, their hand, and I took them by the wrist and hooked their arm underneath my arm.'

Solicitor General: 'You have not told us anything yet about the preference being given to women?'

Lightoller: 'The order had been received from the commander [Smith].'

Commissioner: 'He [Lightoller] has told us about the order given by the captain [Smith].'

Solicitor General: 'I see.'

Solicitor General: [Lightoller] 'And that is the order you carried out?'

Lightoller: 'Yes.'

Solicitor General: 'And then was [lifeboat] number 6 lowered away?'

Lightoller: '[Lifeboat] number 6 was lowered away.'

Solicitor General: 'Was [life]boat number 6 filled?'

Lightoller: 'It was filled with a reasonable regard to safety. I did not count the people going in.'

Solicitor General: 'But you exercised your judgment about it?'

Lightoller: 'Yes.'

Solicitor General: 'It was filled as much as you thought was safe in the circumstances?'

Lightoller: 'Yes.'

Solicitor General: 'In your judgment is it possible to fill these lifeboats when they are hanging as full as you might fill them when they are water borne?'

Lightoller: 'Most certainly not.'

Commissioner: 'Is that due to the weak construction of the lifeboats or to the insufficiency of the falls?'

Lightoller: 'A brand new fall, I daresay, would have lowered the [life]boats down and carried the weight, but it would hardly be considered a seamanlike proceeding as far as the sailor side of it goes, but I certainly should not think that the lifeboats would carry it without some structural damage being done – buckling, or something like that.'

Commissioner: 'And had you those considerations in mind in deciding how many people should go in the [life]boat?'

Lightoller: 'Yes.'

Solicitor General: 'The convenient thing, My Lord, is just to refer your Lordship to the evidence of [Able Seaman John 'Jack' Thomas] Poingdestre [born 1884]. It is on page 83. It fits together here. Perhaps I may read a few questions, and Mr Lightoller will hear them. It is question 2958. He is asked: "Do you know how it comes that there were not more than 42 put into this [life]boat?" That is [life]boat number 6?'

Solicitor General: 'And he [Poingdestre] says: 'Well, the reason is that the falls would not carry any more". [Questioner]: "You mean somebody was frightened of the falls?" [Poingdestre]: "Yes, the second officer Mr Lightoller."'

Solicitor General: 'Did you say anything aloud about it?'

Lightoller: 'No'

Solicitor General: 'It Is merely a conclusion the man comes to?'

Lightoller: 'Yes, I daresay, a seamanlike conclusion.'

Solicitor General: 'You agree as many people were put into it as, in your judgment, was safe when it was in that position?'

Lightoller: 'Yes.'

Solicitor General: 'We are told about 40 or 42?'

Lightoller: 'Yes, about that.'

Solicitor General: 'Then did you give the order to lower away?'

Lightoller: 'Yes.'

Solicitor General: 'Did you give any further order to that [life]boat, number 6, as to what it was to do or where it was to go?'

Lightoller: 'Not that I remember. I knew there was, if I may mention it, this light on the port bow about two points; I had already been calling many of the passengers' attention to it, pointing it out to them and saying there was a ship over there, that probably it was a sailing ship as she did not appear to come any closer, and that at daylight very likely a breeze would spring up and she would come in and pick us up out of the [life]boats, and generally reassuring them by pointing out the light; but whether I told them to pull towards the light I really could not say. I might have done and I might not.'

Solicitor General: 'Here is a [life]boat with only 42 people in it, and when it is water-borne everybody agrees it would safely carry more then?'

Lightoller: 'Yes.'

Solicitor General: 'Did you give any orders with the object of getting more people into it when it was in the water?'

Lightoller: 'Yes, I see what you are alluding to now, the gangway doors. I had already sent the boatswain and 6 men or told the boatswain to go down below and take some men with him and open the gangway doors with the intention of sending the [life]boats to the gangway doors to be filled up. So with those considerations in mind I certainly should not have sent the [life]boats away.'

Solicitor General: 'That is what I meant. Did you give any order or direction to the man in charge of [life]boat number 6 that he was to keep near or was to go to the gangway doors?'

Lightoller: 'Not that I remember. The [life]boats would naturally remain within hail.'

Solicitor General: 'You do not recollect whether you gave any actual order to the man in charge?'

Lightoller: 'No.'

Solicitor General: 'It is just as well to read this question and answer. This man Poingdestre was asked, "Did Mr Lightoller give you any orders as to what to do with the [life]boat?"; and the answer was, "He gave me orders before the [life]boat was lowered what to do".

(Q) "What orders did he give you?"
(A) "To lay off and stand by close to the ship"?'

Lightoller: 'Perhaps I did; I daresay.'

Solicitor General: 'Now let us pursue the two things you have mentioned. You say you gave those orders to the boatswain to go down with some men and open the gangway doors?'

Lightoller: 'Yes.'

Solicitor General: 'Will you point out on the starboard side where they are?'

Lightoller: 'There are gangway doors one on each side there.'

Lightoller: [Pointing on the model.]

Solicitor General: 'About where you are pointing now?'

Lightoller: 'Yes, there are two doors one above and one below on the starboard side, but there is only one on E Deck on the port side. The other gangway doors are here.'

Solicitor General: 'In the afterpart?'

Lightoller: 'Yes.'

Solicitor General: 'What deck do those gangway doors open from?'

Lightoller: 'E Deck.'

Solicitor General: 'Were your orders general, or did they refer to one set of gangway doors in particular?'

Lightoller: 'General.'

Solicitor General: 'Did the boatswain go off after receiving the orders?'

Lightoller: 'As far as I know, he went down.'

Commissioner: 'Have we heard anything up to this time of these gangway doors.'

Solicitor General: 'I am not aware of having heard it, My Lord. There has been a suggestion made by a witness, I think, that it was so, but I do not think there has been any evidence about it. There was a suggestion, I know.'

Commissioner: 'To open those doors?'

Solicitor General: 'Yes.'

Solicitor General: [Lightoller] 'Can you help us when it was that you gave this order to the boatswain? I mean, can you give it us by reference to [life]boats. Was it before you had lowered [lifeboat] number 4 to the A Deck or after?'

Lightoller: 'I think it was after and whilst I was working at number 6 [life]boat.'

Solicitor General: 'If the [life]boat was down by the head, the opening of those doors on E Deck in the forward part of the ship would open her very close to the water, would it not?'

Lightoller: 'Yes.'

Solicitor General: 'When you gave the order, had you got in mind that the ship was tending to go down by the head, or had not you yet noticed it?'

Lightoller: 'I cannot say that I had noticed it particularly.'

Solicitor General: 'Of course, you know now the water was rising up to E Deck?'

Lightoller: 'Yes, of course it was.'

Solicitor General: 'Did the boatswain execute those orders?'

Lightoller: 'That I could not say. He merely said "Aye, aye, sir," and went off.'

Solicitor General: 'Did not you see him again?'

Lightoller: 'Never.'

Solicitor General: 'And did not you ever have any report as to whether he had executed the order?'

Lightoller: 'No.'

Solicitor General: 'I had better just put it. As far as you know, were any of those gangway doors open at any time?'

Lightoller: 'That I could not say. I do not think it likely, because it is most probable the [life]boats lying off the ship would have noticed the gangway doors, had they succeeded in opening them.'

Solicitor General: 'You say you gave that order, as far as you recollect, when you were dealing with that [life]boat number 6?'

Lightoller: 'Yes, [life]boat number 6.'

Commissioner: 'I have the reference now. It is in the evidence of [Archie] Jewell on page 18, questions 131 and 132.'

Solicitor General: 'Yes, My Lord: "What were the orders about – what was she to do?" He speaks of Mr Murdoch giving orders. "He" – that is, Mr Murdoch – "told us to stand by the gangway."'

Commissioner: 'He says this door is open continually. He goes on to say this. The question is put to him – I do not know who was examining him.'

Solicitor General: 'I was I think, My Lord.'

Commissioner: 'Amidships, and the answer is yes.
(Q) "Where the gangway would be if she were in port, I suppose?"
(A) "Yes, that is right."
If this witness is right, he does not seem to know where the gangway was.'

Sir Robert Finlay: 'In the next question he points it out.'

Solicitor General: 'Your Lordship asked him to go to the model.'

Commissioner: 'Just go to the model again and show me where about on that model the waterline was, and where the gangway was, so that I may know where the [life]boat was, and then he indicates. "There is one door there, and there is the waterline right along here. There are several gangway doors in the side; there is one about there somewhere, and one about there." That, of course, tells me nothing, and I do not remember where he pointed. I am told that he pointed further abaft the point indicated by Mr Lightoller.'

Solicitor General: 'I see Mr Wilding here; no doubt he will tell us where, in fact, they are, if your Lordship would like it now.'

Commissioner: 'It occurred to me that Mr Lightoller was right, because I see the rows of portholes?'

Edward Wilding: 'There are the gangway doors here.'

Wilding: [Pointing on the model.]

Commissioner: 'If you look you will see the row of portholes is interrupted.'

Solicitor General: 'Yes. Is that the place where you are pointing now, Mr Wilding?'

Wilding: 'Yes, there is a door marked there.'

Commissioner: 'Is it marked there on the model?'

Wilding: 'Yes, My Lord, here.'

Wilding: [Pointing on the model.]

Solicitor General: 'As a matter of accuracy, is that open on to the floor of E Deck or D deck?'

Wilding: 'E Deck.'

Sir Robert: 'I am told there is also a gangway on D Deck forward?'

Wilding: 'On the starboard side.'

Commissioner: 'On D Deck.'

Sir Robert: 'Yes.'

Solicitor General: 'One above and one below.'

Sir Robert: 'That is on the starboard side forward.'

Commissioner: 'It appears to me that you would be very unlikely to order the forward gangway door to be opened. You might get the head so deep in the water that she might ship water through that gangway door?'

Lightoller: 'Of course, My Lord, I did not take that into consideration at that time; there was not time to take all these particulars into mind. In the first place, at this time I did not think the ship was going down.'

Commissioner: 'I remember what you said yesterday as to what you were told when you were in your bunk that the water was up to F Deck; you knew that it was a very serious state of things?'

Lightoller: 'Yes, I knew it was serious.'

Commissioner: 'And I suppose you realised – I do not know whether you did – but I suppose you realised that the ship was taking in more and more water as you were attending to these [life]boats?'

Lightoller: 'Yes, My Lord, and yet I did not think at that time that the ship was going down.'

Solicitor General: 'Just to get [life]boat number 6 right. The quartermaster, whose name was [Robert] Hichens, was in that number 6 [life]boat. Your Lordship will find a reference to him at page 43; he confirms exactly, of course, what Mr Lightoller is saying. [Solicitor General]: "Who ordered you to another [life]boat?" – [Hichens]: "Mr Lightoller"
[Solicitor General]: "And to what [life]boat"?
[Hichens]: "number 6 [life]boat"
[Solicitor General]: "Is that a lifeboat on the port side?" – [Hichens]: "Yes"

[Solicitor General]: "It would be the third on the port side from forward, would it not?"
– and he says it was the second or third boat to be lowered on the port side.'
Solicitor General: [To Lightoller] 'I understand from you it was the second because you had lowered [life]boat number 4 to A Deck?'

Lightoller: 'Yes.'

Commissioner: 'What question is it?'

Solicitor General: 'I was looking at question 1089. He says it was the second or third [life]boat, and it appears it was really the second.
[Question] 1096 "She had only been swung out ready?"
(A) "That is all."
(Q) "And then what happened – who was giving orders then?" (A) "Mr Lightoller was in charge of the port side."
(Q) "Did you hear any order given?"
(A) "Yes, I heard the captain say 'Women and children first'."
[Question] 1106 – "How many people did you take on board?" (A) "42, all told".'

Solicitor General: [To Lightoller] 'I think a gentleman named Major [Arthur Godfrey] Peuchen [1859–1929] was one of them?'

Lightoller: 'Yes.'

Day Fourteen, 3 May (re-called)

Sir Robert Finlay: 'You have heard that a message was sent, according to the evidence, to the *Titanic* for transmission to Cape Race [Marconi Station] from the [SS] *Amerika*?'

Lightoller: 'Yes.'

Sir Robert: 'Which would reach *Titanic* about 2 pm?'

Lightoller: 'Yes.'

Sir Robert: 'You know the nature of that message?'

Lightoller: 'I heard it, yes.'

Sir Robert: 'And that a message is said to have been sent from the *Mesaba* which could not reach the *Titanic* before about 10 pm?'

Lightoller: 'Yes.'

Sir Robert: 'You have heard that?'

Lightoller: 'Yes, I have heard of that also.'

Sir Robert: 'Did you ever hear of any such messages?'

Lightoller: 'Nothing whatever.'

Sir Robert: 'What was the course of business with regard to messages which are communicated by the Marconi operators to the captain or officers?'

Lightoller: 'It is customary for the message to be sent direct to the bridge. If addressed "The Captain," or "Captain Smith," it is delivered to Captain Smith personally, if he was in the quarters or about the bridge. If Captain Smith is not immediately gettable, if not in his room or on the bridge, it is then delivered to the senior officer of the watch. Captain Smith's instructions were to open all telegrams and act on your own discretion.'

Sir Robert: 'And are you positive that you never heard anything of either of those telegrams?'

Lightoller: 'Absolutely positive.'

Sir Robert: 'What were you doing during the day; just recapitulate in this connection what you were doing. In the afternoon, about 2 o'clock, where would you be?'

Lightoller: 'I was below.'

Sir Robert: 'When did you come up?'

Lightoller: 'At 6 o'clock.'

Sir Robert: 'And from 6?'

Lightoller: 'From 6 till 10, with the exception of half an hour for dinner.'

Sir Robert: 'You were on the bridge?'

Lightoller: 'I was.'

Sir Robert: 'And nothing was said by anyone about such telegrams?'

Lightoller: 'There was no telegram received by me nor did I hear of any telegram.'

Sir Robert: 'Were you in communication with the captain [Smith] and with other officers during that time?'

Lightoller: 'Between six and ten?'

Sir Robert: 'Yes?'

Lightoller: 'I was in communication with the chief officer [Wilde] when I relieved him, and with the first officer [Murdoch] when I was relieved by him for dinner, and with the commander [Smith] when he was on the bridge, as well as junior officers.'

Sir Robert: 'How often, and for how long, did you see the commander [Smith] on the bridge?'

Lightoller: 'He came on the bridge about five minutes to 9, and remained with me till about twenty or twenty-five past nine.'

Sir Robert: 'A message such as that from the "*Mesaba*" would be one, of course, of great importance?'

Lightoller: 'I have no doubt it would have been immediately communicated to me if it referred to pack ice, as I believe it does.'

Solicitor General: 'May I ask him a question or two about it?'

Solicitor General: 'How many messages about ice on the 14th [April 1912] have you any knowledge of?'

Lightoller: 'I have a distinct recollection of the message that the commander [Smith] brought on the bridge to me, and which I mentioned as having read while he held it in his hands.'

Solicitor General: 'You told us that was about a quarter to one?'

Lightoller: 'Yes.'

Solicitor General: 'I will give your Lordship the reference, if I may. That will be found in this witness's evidence at page 302, question 13466. Perhaps I may read two or three before that. Your Lordship had asked: "What time was it?" and I had said: "So far, My Lord, he has said it was between 12.30 and 1 pm in the middle of the day"; and then I said to Mr Lightoller: "(13,460) Can you fix at all as between those times?"
(A) "About 12.45 as near as I can remember."
(Q) "Very well; about a quarter to 1?"
(A) "Yes." [Laing] "I have the wording of it,"
and he handed to me the wording of the *Caronia* message. I read that to the witness. Then I said at Question 13463: "You had not heard anything about that before you went off your watch at 10 o'clock?"
(A) "No."
(Q) "Can you help us? Would 9.44 am. *Caronia*'s time coming from New York be likely to be later than your 10 o'clock watch coming to an end? You see, you went off duty at 10?"
(A) "Yes."
[Commissioner] "Did Captain Smith tell you when he had received the Marconigram?"
(A) "No, my Lord."
[Solicitor General] "And the first you knew of it was when Captain Smith showed it to you at about a quarter to one?"
(A) "Yes."
(Q) "So far as your knowledge goes, is that the first information as to ice which you had heard of as being received by the '*Titanic*'?"
(A) "That is the first I have any recollection of. That is that one."'

Commissioner: 'Where is that last question?'

Solicitor General: 'The very bottom question on page 302.'

Solicitor General: [To Lightoller] 'That is the *"Caronia's"* message, so that we may fairly treat that as identified and brought to your notice in that way?'

Lightoller: 'Yes.'

Solicitor General: 'Now apart from that message, were not other messages, in your belief, received to the knowledge of the officers about ice on the 14th [April 1912]?'

Lightoller: 'To my belief there were perhaps some messages, but I can give no information and I cannot recollect with any degree of distinctness having seen them.'

Solicitor General: 'I will tell you why I put the question, and I think My Lord will remember it. I put it to you for this reason. I asked you if you recollected when you were here the other day, whether Mr Moody, when he calculated that you would reach the ice at 11 pm., had, you thought, used the *"Caronia"* message, and you told me your impression was he had used another message; is not that so?'

Lightoller: 'Precisely.'

Solicitor General: 'That is in the middle of page 304, question 13,531. You will see the answer: "I directed the sixth officer [Moody] to let me know at what time we should reach the vicinity of the ice. The junior officer reported to me, 'About 11 o'clock.'"
(Q) "Do you recollect which of the junior officers it was?"
(A) "Yes, Mr Moody, the sixth" That would involve his making some calculations, of course?"
(A) "Yes."
(Q) "Had this Marconigram about the ice, with the meridians on it, been put up; was it on any notice board, or anything of the sort?"
(A) "That I could not say with any degree of certainty. Most probably, in fact very probably, almost certainly, it would be placed on the notice board for that purpose in the chart room." (Q) "At any rate, when you gave Mr Moody those directions he had the material to work on?"
(A) "Exactly."
(Q) "And he calculated and told you about 11 o'clock you would be near the ice?"

(A) "Yes." Then the next question and answer: "That is to say an hour after your watch finished?"
(A) "Yes. I might say, as a matter of fact, I have come to the conclusion that Mr Moody did not take the same Marconigram which Captain Smith had shown me on the bridge, because, on running it up just mentally, I came to the conclusion that we should be to the ice before 11 o'clock by the Marconigram that I saw." Then your Lordship says: "In your opinion, when, in point of fact, would you have reached the vicinity of the ice?"
(A) "I roughly figured out about half-past 9."
(Q) "Then had Moody made a mistake?"
(A) "I should not say a mistake, only he probably had not noticed the 49 degree wireless"
– that is the "*Caronia*" one you had seen?'

Lightoller: 'Yes.'

Solicitor General: 'There may have been others, and he may have made his calculations from one of the other Marconigrams. (Q) "Do you know which other Marconigram he would have to work from?"
(A) "No, My Lord, I have no distinct recollection of any other Marconigrams."
(Q) "Because it is suggested to me that there was no Marconigram which would indicate arrival at the ice-field at 11 o'clock?"
(A) "Well, My Lord, as far as my recollection carries me, Mr Moody told me 11, and I came to that conclusion that he had probably used some other Marconigram?"'

Lightoller: 'Exactly.'

Solicitor General: 'As a matter of fact, if one takes the Marconigram, for instance, from the "Baltic," which we proved today, it would give a later time than 9.30, and it would bring you to something like 11 o'clock. Have you noticed that?'

Lightoller: 'No, I have not. I think it will be found so.'

Sir Robert: 'Which message?'

Solicitor General: 'I am calling his attention to the circumstances. The "*Caronia*" message mentioned you getting to ice as soon as you got to the 49th meridian?'

Lightoller: 'Exactly.'

Solicitor General: 'I do not like to make a suggestion unless the Admiral thinks it is correct, but I think that is substantially so.'

Solicitor General: [To Lightoller] 'You see what I mean?'

Lightoller: 'Yes.'

Solicitor General: 'And your impression at the time was not that Mr Moody had made a mistake in his calculations, but that he had used another Marconigram?'

Lightoller: 'Exactly. You will quite understand that all I am quoting is purely from memory. I am trying as much as I possibly can, of course, to assist, and it is just these mere facts as I recollect them with regard to 11 o'clock. There is nothing to identify 11 o'clock in my mind. Merely what I recollect, and also with regard to the Marconigrams. I put that down as the most feasible explanation of the 11 o'clock, but I cannot say, of course, that Mr Moody actually had seen other Marconigrams.'

Solicitor General: 'Oh, no; you have been perfectly fair and candid about it, as far as I am concerned, if I may say so. You did not ask Mr Moody to make the calculation again or check it?'

Lightoller: 'No.'

Solicitor General: 'You accepted his [Moody's] statement that his calculation showed 11 o'clock?'

Lightoller: 'Yes.'

Solicitor General: 'I think one also ought to put it from this point of view. Let me take these telegrams in order and see which of them would come in your watch, as far as one can judge. Your watch was from 6 to 10 am and 6 to 10 pm?'

Lightoller: 'Exactly.'

Solicitor General: 'You also, I think, relieved Mr Murdoch, you told us, between half past 12 and 1, at lunchtime?'

Lightoller: 'Yes.'

Solicitor General: 'And during your evening watch, from six to ten you were off for a certain time to dinner?'

Lightoller: 'Exactly.'

Solicitor General: 'Now those are the times for which you are responsible. The *"Caronia"* message by your ship's time would get to your ship about 11 o'clock.'

Commissioner: 'Ship's time?'

Solicitor General: 'Yes.'

Solicitor General: [To Lightoller] 'Or something of the sort?'

Lightoller: 'Yes.'

Solicitor General: 'It was acknowledged at 9.44 [am], New York time, and, adding two hours, will make it between 11 and 12?'

Lightoller: 'Yes.'

Solicitor General: 'You would have finished your morning watch by then?'

Lightoller: 'I should.'

Solicitor General: 'And you would be off duty?'

Lightoller: 'Yes. I may incidentally mention the fact that I should be on the bridge between a quarter to 12 and a minute or two past 12 taking the noon position; I should be there with the commander [Smith] and the chief [Wilde] and first [Murdoch] officers.'

Solicitor General: 'But at any rate you did not hear anything of the *"Caronia"* message at that time?'

Lightoller: 'Nothing.'

Solicitor General: 'You did hear of the *"Caronia"* message at about a quarter to one, when you were relieving Mr Murdoch while he had lunch?'

Lightoller: 'About that time, yes.'

Solicitor General: 'The next message in order of time that is suggested is the *"Amerika"* message, which merely goes through the *"Titanic"*?'

Lightoller: 'Yes.'

Solicitor General: 'And that would go through apparently about 2 o'clock?'

Sir Robert: 'No; it ought to have been received about 2 [pm], but it could not go on till 8.30 [pm]. It would be put up with other messages and transmitted after 8.30 [pm] to Cape Race.'

Solicitor General: 'You are quite right. It would be in the custody of the Marconi room at some time about 2 [pm], and presumably would be kept until they got into communication with Cape Race.'

Sir Robert: 'Yes.'

Solicitor General: 'When it arrived you were off duty. Assuming this evidence is right it would be in the Marconi room at 6 o'clock when you came on duty again?'

Lightoller: 'Yes.'

Solicitor General: 'You heard nothing of that?'

Lightoller: 'Nothing.'

Solicitor General: 'The next one is the message from the *"Baltic"* which, as I pointed out just now, would give the position of the ice at about 11 pm?'

Lightoller: 'Yes.'

Solicitor General: 'That message from the *"Baltic"* would get to your ship at about 1 o'clock?'

Lightoller: 'I think so, 1 pm.'

Solicitor General: 'You would be off duty?'

Lightoller: 'Yes.'

Solicitor General: 'Do you observe that if you told Mr Moody when you came on duty at 6 pm to calculate when he would meet ice, the "*Baltic*" message would be a later message in point of time than the "*Caronia*" message?'

Lightoller: 'I see.'

Solicitor General: 'Then the '*Californian*' message?'

Lightoller: 'If I may interrupt you to make it a little clearer; when I gave Mr Moody instructions (I think if I did not say it in my evidence, I ought to have done.) I used words to the effect that would guide him to look for the earliest ice, to let me know at what time we should be up at the ice. He would naturally look at the easternmost.'

Solicitor General: 'When you gave him instructions, as far as you knew there was only one ice message?'

Lightoller: ''Yes.

Solicitor General: 'You did not know of two?'

Lightoller: 'No.'

Solicitor General: 'Then if I take the "*Californian*" message, it appears that that message passed at about half past 7, ship's time. That is right, Sir Robert, I think.'

Sir Robert: 'Yes. Of course, there is a conflict between the *procès-verbal* and the other witness.'

Solicitor General: [To Lightoller] 'You were on duty between 6 and 10?'

Lightoller: 'Yes.'

Solicitor General: 'So that that message, if it arrived at 7.30 [pm], would arrive during your evening watch?'

Lightoller: 'Yes.'

Solicitor General: 'But you are off duty at some time between 6 and 10 [pm], in order to get dinner?'

Lightoller: 'Yes.'

Solicitor General: 'What is the sort of time you are off duty?'

Lightoller: Half an hour. I think that is 7.05 to 7.35 [pm], as near as I remember.'

Solicitor General: 'And who took your place when you were off duty?'

Lightoller: 'Mr Murdoch, the first officer.'

Solicitor General: 'You knew nothing of the *"Californian"* message at all?'

Lightoller: 'Nothing whatever.'

Solicitor General: 'Then the last one, the *"Mesaba"* message, according to the evidence given, would reach your ship about 10 o'clock?'

Lightoller: 'Yes.'

Solicitor General: 'That is when you would be changing watch, and Mr Murdoch would be taking your place?'

Lightoller: 'Yes.'

Commissioner: 'You told Mr Moody you wanted him to ascertain the time when you would meet the most easterly of the ice. Was that so?'

Lightoller: 'That is the impression I wished to convey, whether I actually used the word easterly I do not recollect, but he would naturally conclude that, I should judge.'

Commissioner: 'The information in the "*Caronia*'s" telegram would indicate that the ice there referred to was considerably to the north of the track?'

Lightoller: 'I believe so.'

Commissioner: 'Is it possible that Mr Moody may have calculated the position of the ice given by the "*Baltic*'s" telegram?'

Lightoller: 'It is possible, but it is most probable that he would pay the greatest attention to the longitude regardless of the latitude.'

Commissioner: 'But if he did calculate according to the '*Baltic*'s' telegram, he would ascertain the time at which the ice would be arrived at as 11 o'clock?'

Lightoller: 'Quite so.'

Commissioner: 'And the "*Baltic*'s" information was to the effect that ice was on the track?'

Lightoller: 'A little to the north.'

Solicitor General: 'If your Lordship will turn to page 366, Mr Lowe's evidence, you will see why I think it well to put it to this gentleman.'

Solicitor General: [To Lightoller] 'Let me tell you how the matter stands. You are on duty from 6 to 10 in the evening and about half past seven according to the "*Californian*" witnesses, there was a message sent from the "*Californian*," of which you know nothing?'

Lightoller: 'That is right.'

Solicitor General: 'You, as a matter of fact, were off for dinner for half an hour from seven to half past?'

Lightoller: 'Yes.'

Solicitor General: 'I am referring to the questions beginning 15778. Did you see anything at all of a piece of paper, not in an envelope – a small piece of paper – a square chit of paper about 3 by 3 with the word 'ice' on it any time between 6 and 8?'

Lightoller: 'No.'

Solicitor General: 'What would be meant by seeing a small piece of paper on the chart room table? Which room is it?'

Lightoller: 'Leading out of the wheelhouse on the afterpart of the port side.'

Solicitor General: 'It is the thing which is marked on my plan as the chart house then?'

Lightoller: 'Yes.'
Solicitor General: 'Is there a table there?'

Lightoller: 'There is.'

Solicitor General: 'And supposing there is a message about ice and it cannot be given personally to the captain [Smith], where would such a message be put?'

Lightoller: 'It would not be put anywhere; it would be brought out on the bridge to the senior officer of the watch.'

Solicitor General: 'Whoever he was?'

Lightoller: 'Whoever he was.'

Solicitor General: 'This little room, the chart house, is immediately aft of the wheel house?'

Lightoller: 'On the port side, yes.'

Solicitor General: 'You heard nothing of that?'

Lightoller: 'Nothing.'

Solicitor General: 'And you were off for dinner for half an hour?'

Lightoller: 'Yes.'

Commissioner: 'Why would the piece of paper with the word "ice" upon it be placed there?'

Lightoller: 'I may say I do not quite follow what you mean by the word "ice" unless you are alluding to a message written on a chit of paper.'

Commissioner: 'This is the evidence. He is asked on page 366, question 15779: "You were on duty from 6 to 8?"
(A) "I was."
(Q) "Did you hear anything about any messages about ice?"
(A) "There was a chit on the chart room table with the word "ice" on – meaning 'ice' on the piece of paper."'

Solicitor General: 'Will your Lordship read the next two or three questions.'

Commissioner: 'Yes. "You mean a little piece of paper with 'ice' written on it?"
(A) "A square chit of paper about 3 by 3."
(Q) "On the chart room table?"
(A) "On our chart room table."
(Q) "What is that, 'Our chart room table'"?
(A) "The officers' chart room table, and the word "ice" was written on top and then a position underneath."
(Q) "Can you remember what the position was?"
(A) "I cannot. What is this chart room table?"'

Lightoller: 'It consists of the top of a chest of drawers. In those drawers are all the charts, necessarily big drawers, to contain the charts fully laid out, and also drawers for navigational books, instruction books, and so on.'

Commissioner: 'Would that chit of paper be placed there by somebody with the position marked upon it so that a chart might be consulted for the purpose of finding out where that ice was?'

Lightoller: 'A track chart is always lying on that chart room table. I quite understand what a chit of paper is. There are little pads, position pads, and deviation pads, and it is customary to tear off one of these chits and write on the back; and it would have been left on the chart room table, lying on the top of the chart.'

Solicitor General: 'Were you in court here this morning when Mr Bride gave evidence?'

Lightoller: 'I was.'

Solicitor General: 'Did you hear him say that the message heard from the "*Californian*" he wrote down on a bit of paper, but he did not put it in an envelope?'

Lightoller: 'Yes.'

Solicitor General: 'And if the message from the "*Californian*" came at half past 7, then it would be on that watch of Mr Lowe's that he is referring to here, 6 to 8?'

Lightoller: 'Yes.'

Commissioner: 'You knew nothing of that. Are these messages which come from the Marconi room written on chits of paper?'

Lightoller: 'No, My Lord.'

Commissioner: 'They are on forms?'

Lightoller: 'On proper telegraph forms. My explanation of that chit of paper would be that an officer has copied from some wireless telegram; he has noticed that there has been an ice-position on, and he has just scribbled down on a piece of paper "ice," and the position, and then has probably gone to the chart room, found the position, and marked it on the chart, and left the paper there, instead of crumpling it up and throwing it away; but I do not think that chit was of any importance, and I do not think it came from the Marconi room – except, I mean, as a copy of the wireless.'

Solicitor General: 'Do not say it is not of importance. When you say it had a position, you mean it stated probably the latitude and longitude?'

Lightoller: 'Yes.'

Solicitor General: 'Do you know what Mr Lowe says he did about it. Just look at page 370. There is a question asked by Sir Robert Finlay: (Q 15984) "You saw this chit, the note about the ice on the table?"
(A) "Yes."
(Q) "Did you work it out?"

(A) "I worked it out roughly."
(Q) "You were on watch 6 to 8?"
(A) "Yes. I ran this position through my mind, and worked it out mentally and found that the ship would not be within the ice region during my watch, that is from 6 to 8."
(Q) "You do not recollect what the figures were?"
(A) "I do not."
(Q) "But that was the result you arrived at?"
(A) "That was the result I arrived at."'

Sir Robert: 'May I ask one question on that? You have been asked about the instructions you gave as to working out the time when you would get to the ice?'

Lightoller: 'Yes.'

Sir Robert: 'About what time was it you gave those instructions?'

Lightoller: 'Soon after I came on deck. That is, soon after 6 o'clock.'

Sir Robert: 'And when did you get the report?'

Lightoller: 'It was some time later, because they were working stars; probably shortly before 7 o'clock.'

Sir Robert: 'That, of course, was long before any *"Mesaba"* message could, by any possibility, have reached the *"Titanic"*?'

Lightoller: 'Yes, I believe so.'

Sir Robert: 'You have heard the *"Mesaba"* message, of course?'

Lightoller: 'Yes.'

Sir Robert: 'Is that a message which, if the captain or any officer had got, he could have failed to communicate to his colleagues?

Lightoller: I think had that message been delivered, even to the captain [Smith], he would immediately have brought the message out personally to the bridge; he would not even have sent it out, and he would have seen it was communicated to all the senior officers, as well as distinctly marked on the chart. It was of the utmost importance.'

Sir Robert: 'And of a somewhat startling character?'

Lightoller: 'Extremely so.'

Sir Robert: 'The captain [Smith], I think you said, had been on the bridge at 9.30 [pm]?'

Lightoller: 'From 5 minutes to 9 till 20 or 25 minutes past.'

Commissioner: 'Will you tell me what messages, to begin with, about ice you saw on the 14th [April 1912]?'

Lightoller: 'The one that the commander [Smith] brought on to the bridge in his own hands to me shortly after midday.'

Commissioner: 'Is that the "*Caronia*"?'

Lightoller: 'I believe that is the "*Caronia*'s" message.'

Commissioner: 'Now, did you see any other message about ice?'

Lightoller: 'I cannot give any distinct recollection of having seen any other. You will quite understand we are in and out of the chart room, and I may have seen notices on the board. If they were there I should read them.'

Commissioner: 'I am talking about messages from the Marconi room. Would they be pinned up on a board?'

Lightoller: 'Yes.'

Commissioner: 'You do not remember seeing any other than the "*Caronia*'s"?'

Lightoller: 'That is what I am explaining. If they were pinned up on this board and I was in the chart room – which we are frequently – I should notice them, make a mental note of the position of the ice, take the most easterly position, and then disregard the rest.'

Commissioner: 'That is to enable you to ascertain how soon you may expect to reach the ice?'

Lightoller: 'Exactly.'

Commissioner: 'Can you tell me what other ice message besides the "*Caronia*'s" you heard of?'

Lightoller: 'I heard of none that I remember.'

Commissioner: 'Did you hear any conversation about any other ice message?'

Lightoller: 'None.'

Commissioner: 'You did not hear anything about the "*Californian*'s" ice message?'

Lightoller: 'Of no message except that one I spoke of from 49 deg. [degrees], to 51 deg. [degrees].'

Commissioner: 'I daresay you have in your mind the messages which have been referred to?'

Lightoller: 'Yes.'

Commissioner: 'The ice mentioned in the "*Caronia*'s" message was the easternmost ice of all, was it not?'

Lightoller: 'I believe so.'

Commissioner: 'Now, it is suggested that as you would want to know the most easterly ice you may have disregarded the other messages which indicated ice further west, and may only have bent your mind upon the most easterly ice. Do you think that is so?'

Lightoller: 'Exactly, My Lord, with this reservation that had there been any mention of pack ice there is no doubt I should have fixed that telegram in my mind.'

Third Officer Herbert John Pitman

Herbert Pitman was questioned in the United States on 20 April 1912, and by the British inquiry on days thirteen and fourteen as follows:

Day Thirteen, 22 May 1912

Butler Aspinall: 'Were you serving as third officer on the *Titanic* at the time of this accident?'

Pitman: 'Yes.'

Aspinall: 'What certificate do you hold?'

Pitman: 'Ordinary master.'

Aspinall: 'How long have you been in the service of the White Star Line?'

Pitman: 'Five and a half years.'

Aspinall: 'And during those five and a half years have you been travelling backwards and forwards across the Atlantic?'

Pitman: 'Twelve month only.'

Aspinall: 'And have you had considerable experience on the sea in other parts of the world besides the Atlantic?'

Pitman: 'Sixteen years.'

Aspinall: 'Twelve months experience in the Atlantic?'

Pitman: 'Of the North Atlantic.'

Aspinall: 'I will get this fact from you now, it comes a little later in your story. You were saved in [life]boat 5 were you not?'

Pitman: 'Yes.'

Aspinall: 'We have had some evidence with regard to [life]boat 5, My Lord, but not of a very satisfactory character. It is the evidence of [Alfred Charles] Shiers [1886–1946], the fireman. He was not able

to give us very useful evidence with regard to the matter: [read out by Aspinall] "Question: Then you got into [lifeboat] number 5? Was [lifeboat] number 5 lowered?"
Shiers: "Yes."
Question: "Who were in [lifeboat] number 5? You were and who else?"
Shiers: "One other fireman, a steward and a quartermaster."
Question: "And were there some women and children in number 5?"
Shiers: "Women, no children."
Question: "Do you know how many women?"
Shiers: "No."
Question: "Was the boat full or not?"
Shiers: "It was not full – as many as would take off the davits was what the officer said – as many as he thought the [life]boat would take off the davits."'

Aspinall [| To Pitman.]: 'To come to Sunday, 14 April [1912]. At the time of the accident you were off watch and asleep in your cabin, were you not?'

Pitman: 'That is right.'

Aspinall: 'When before, on that day, had you last been on duty?'

Pitman: '6 to 8 pm.'

Aspinall: 'And before in the course of that day, had you been on duty?'

Pitman: '12 to 4 in the afternoon.'

Aspinall: 'Was it within your knowledge that the ship would probably meet with ice that evening?'

Pitman: 'We knew that we would be in the longitude of ice.'

Aspinall: 'Who told you that?'

Pitman: 'I saw it in a Marconigram.'

Aspinall: 'Do you know from which ship the Marconigram had come?'

Pitman: 'I have no idea.'

Aspinall: 'Did you particularly concern yourself with that matter?'

Pitman: 'No, I simply looked at them and saw that there was no ice reported on the track.'

Aspinall: 'Did you see one Marconigram or two Marconigrams?'

Pitman: 'Two, I think.'

Aspinall: 'Were these posted in some part of the ship?'

Pitman: 'Yes, in the chart room.'

Aspinall: 'And you read them, did you?'

Pitman: 'Yes.'

Aspinall: 'Was there any discussion between you and any of the other officers about the fact that you would probably meet ice that night?'

Pitman: 'I do not think so.'

Aspinall: 'Do you mean you do not remember?'

Pitman: 'I do not remember.'

Aspinall: 'As you have said, you were turned in at the time the vessel struck the iceberg, and, I believe, asleep?'

Pitman: 'Yes.'

Aspinall: 'You were aroused and at first did you think much had happened?'

Pitman: 'No, I did not.'

Aspinall: 'What was it aroused you? Was it a noise, or a jar, or what?'

Pitman: 'A noise. I thought the ship was coming to anchor.'

Aspinall: 'Did you lie in your bunk for a few minutes.'

Pitman: 'I did.'

Aspinall: 'At the end of those few minutes did you do anything?'

Pitman: 'Yes, I went on deck.'

Aspinall: 'Was that curiousity, or what took you there?'

Pitman: 'Yes, I suppose it was.'

Aspinall: 'Getting on deck, what did you see or hear?'

Pitman: 'I saw nothing and heard nothing.'

Aspinall: 'Did you go to the forward part of the navigation bridge?'

Pitman: 'No, I only just went outside the [officers'] quarters.'

Aspinall: 'The officers' quarters?'

Pitman: 'That is all.'

Aspinall: 'As it were, put your head out and saw nothing?'

Pitman: 'No, I went on deck.'

Aspinall: 'Seeing and hearing nothing, what did you do then?'

Pitman: 'I went back inside again.'

Aspinall: 'And turned in again?'

Pitman: 'No, I met Mr Lightoller first of all, and I asked him what had happened, if we had hit something, and he said, "Yes, evidently."'

Aspinall: 'He said, "Evidently"?'

Pitman: 'Yes, evidently something had happened.'

Aspinall: 'After you had received that information what did you do?'

Pitman: 'I went to bed.'

Aspinall: 'How long did you remain in bed?'

Pitman: 'It must have been five minutes.'

Aspinall: 'And at the end of five minutes what did you see?'

Pitman: 'I thought I might as well go up, as it was no use trying to go to sleep again, as I was due on watch in a few minutes.'

Aspinall: 'Your watch was the middle watch from 12 to 4 am?'

Pitman: 'That night, yes.'

Aspinall: 'Did you get up and proceed to dress?'

Pitman: 'Yes.'

Aspinall: 'While you were dressing did you receive any information?'

Pitman: 'Mr Boxhall came to my room and said the mail room was afloat.'

Aspinall: 'How long do you think had elapsed between the time you were aroused and Mr Boxhall coming and telling you this?'

Pitman: 'I should think it must be 20 minutes.'

Aspinall: 'Did he give you any information as to what had caused the mail room to be afloat?'

Pitman: 'Yes, I asked him what we had struck, and he said "an iceberg".'

Aspinall: 'After that did you quickly proceed with your dressing?'

Pitman: 'Yes, I put my coat on and went on deck.'

Aspinall: 'When you got on deck, did you see anything being done?'

Pitman: 'The men were uncovering the [life]boats.'

Aspinall: 'On which side was that?'

Pitman: 'That was the port side.'

Aspinall: 'Did you meet the sixth officer when you went on deck, Mr Moody?'

Pitman: 'Yes, I met him on the afterpart of the deck.'

Aspinall: 'Did he give you any information?'

Pitman: 'No, I asked him if he had seen the iceberg; he answered, "No, but there was ice on the foreward well deck."'

Aspinall: 'I believe you, at the time, did not think anything serious had happened, did you?'

Pitman: 'I did not.'

Aspinall: 'Then, I think, you went and looked at some ice, and, after having looked at the ice, did you then go under the forecastle head to see if any structural damage had been done to the bow of the ship?'

Pitman: 'Exactly.'

Aspinall: 'I believe you saw none. As you were coming from the forecastle, did you see any firemen?'

Pitman: 'Yes, I saw a whole crowd of them coming up from below.'

Aspinall: 'Did you ask them what was causing them to come up?'

Pitman: 'Yes.'

Aspinall: 'What was their answer?'

Pitman: 'That water was coming into their quarters.'

Aspinall: 'From which side were the firemen coming, the port or starboard side?'

Pitman: 'The starboard side.'

Aspinall: 'A consequence of what they told you, did you go and do anything?'

Pitman: 'No, I simply looked down number 1 hatch and saw water rushing up number 1 hatch, or at least round it.'

Aspinall: 'Is that the hatch which has the coamings which I think we were told was on G Deck?'

Pitman: 'Yes, the same one that [Lookout George Thomas Macdonald] Symons [1888–1950] was speaking of the other day.'

Aspinall: 'Was the water coming in fast or slow, or how?'

Pitman: 'Quite a little stream, both sides of the hatch.'

Aspinall: 'Did you notice what direction it was flowing from; was it flowing from forward to aft, or how, or did you not notice?'

Pitman: 'Well, I think it was running mainly from the starboard side.'

Aspinall: 'Running from the starboard side?'

Pitman: 'Yes.'

Commissioner: 'I did not understand that.'

Aspinall: [To Pitman] 'What do you mean by from the starboard side?'

Pitman: 'Coming in from the starboard side of the ship.'

Commissioner: 'That I understand.'

Aspinall: 'Seeing that, did you then go back to the boat deck?'

Pitman: 'Yes.'

Aspinall: 'Were the [life]boats still being uncovered, or had they finished the uncovering of the [life]boats by then?'

Pitman: 'I could not say what had happened on the port side. I then returned to the starboard side and they were still uncovering the [life]boats.'

Aspinall: 'You are now on the starboard side, and I think you remained on the starboard side, did you not?'

Pitman: 'Yes.'

Aspinall: 'Did you see the first officer [Murdoch] taking part in getting the [life]boats ready?'

Pitman: 'No, I did not see him.'

Aspinall: 'Did you hear any orders being given?'

Pitman: 'No more than getting the [life]boats filled with women and children, that is all I heard.'

Aspinall: 'Did you go to any one of these [life]boats?'

Pitman: 'Yes, I went to [lifeboat] number 5.'

Aspinall: 'Did you go to [lifeboat] number 7 first?'

Pitman: 'No, Mr Murdoch was there. I did not see him.'

Aspinall: 'Which was your [life]boat?'

Pitman: '[Emergency lifeboat] number 1 usually in case of emergency.'

Aspinall: 'If there was an emergency you would take charge of [lifeboat] number 1, is that so?'

Pitman: 'Yes, that is the case of a man overboard, and things like that.'

Aspinall: 'Was your name on the [life]boat list, as being the officer to look after that [life]boat?'

Pitman: 'Yes, in an emergency.'

Aspinall: 'We are told that there are [life]boat lists put up about the ship; that is so, is it not?'

Pitman: 'That is so.'

Commissioner: 'Did you ever read your name on any list?'

Pitman: 'I did not, as it is an understood thing that the third officer [Pitman] looks after number 1 [life]boat.'

Commissioner: 'You did not see your name on any list?'

Pitman: 'No.'

Commissioner: 'We have been told that there are [life]boat lists put up about the ship; that is so, is it not?'

Pitman: 'That is so.'

Commissioner: 'Did you ever read your name on any list?'

Pitman: 'I did not, as it is an understood thing the third officer [Pitman] looks after number 1 boat.'

Commissioner: 'You did not see your name on any list?'

Pitman: 'No.'

Aspinall: 'Would it be your duty to inform yourself as to what your [life]boat was according to the list?'

Pitman: 'No, it is quite an understood thing in the company for the third [Pitman] and fourth [Boxhall] officers to have number 1 and number 2 [life]boat[s].'

Aspinall: 'Apart from understandings, would it be your duty at the beginning of the voyage to go and ascertain what [life]boat was your [life]boat?'

Pitman: 'No.'

Aspinall: 'It is not your duty?'

Commissioner: 'If, Mr Aspinall, it was the invariable practice for him to attend to number 1 emergency [life]boat, there was no occasion for him to look at the list.'

Aspinall: 'No, My Lord.'

Aspinall: [To Pitman] 'Now you went in fact to [lifeboat] number 5. Why was that?'

Pitman: 'Mr Murdoch ordered me there.'

Aspinall: 'Was there any other officer there?'

Pitman: 'I did not see anyone.'

Aspinall: 'Of course, you know all the officers?'

Pitman: 'Oh, yes.'

Aspinall: 'If there had been one there you would have known?'

Pitman: 'Mr Murdoch was there before the [life]boat was lowered.'

Aspinall: 'Had you seen Murdoch there at [lifeboat] number 5, or merely heard his voice?'

Pitman: 'Oh, no, I saw him.'

Aspinall: 'At [lifeboat] number 5?'

Pitman: 'At [lifeboat] number 5, after the [life]boat was out and practically filled with passengers.'

Aspinall: 'When you got to [lifeboat] number 5, in what state was [lifeboat] number 5?'

Pitman: 'Well, the cover was still on.'

Aspinall: 'How long do you think had elapsed from the time of striking the [ice]berg up to the time you got to [lifeboat] number 5? It is difficult, I know, to be certain about time. Was it half an hour or 45 minutes? Let me help you. You gave me one space of time – about 20 minutes?'

Pitman: 'Yes, I remember that.'

Aspinall: 'Will that help you to approximate what you think was the time between the striking of the iceberg and you getting to [life]boat number 5? Was it an hour, do you think?'

Pitman: 'No, I should think it would be about 12.20 [am].'

Aspinall: 'You say the cover was still on. Was the cover being stripped at the time you got there?'

Pitman: 'It was being uncovered then – Yes.'

Aspinall: 'Did you see Mr [Joseph Bruce] Ismay close to this [life]boat?'

Pitman: 'I did.'

Aspinall: 'Was he taking any part, saying anything, or doing anything?'

Pitman: 'He remarked to me as we were uncovering the [life]boat, "There is no time to lose." Of course, I did not know who he was then, and therefore did not take any notice.'

Aspinall: 'You have since learned that that gentleman was Mr Ismay, have you?'

Pitman: 'Yes.'

Aspinall: 'How many men had you helping [you] at this [life]boat?'

Pitman: 'I think four.'

Aspinall: 'Were they sailor men, or could you tell in the darkness of the night?'

Pitman: 'Well, I knew that two were.'

Aspinall: 'And was the [life]boat uncovered and swung out?'

Pitman: 'Yes.'

Aspinall: 'What was done with it? Was it then lowered to the level of the boat deck?'

Pitman: 'It was lowered level.'

Aspinall: 'And after you had got out to the level of the boat deck, what did you do with regard to passengers?'

Pitman: 'Mr Ismay remarked to me to get it filled with women and children, to which I replied, "I will await the commander's [Smith's] orders." I then went to the bridge, and I saw Captain Smith, and I told him what Mr Ismay had said. He said, "Carry on."'

Commissioner: 'What does that mean?'

Pitman: 'Go ahead [loading the lifeboats].'

Aspinall: 'At this time, did you realise that this gentleman was Mr Ismay, or did you still think he was one of the passengers?'

Pitman: 'Oh, I knew then that it was Mr Ismay – yes, judging by the descriptions I had had given me of him.'

Aspinall: 'The captain [Smith] told you to "carry on". Did you then return to the boat deck?'

Pitman: 'I was already there; I returned to [lifeboat] number 5.'

Aspinall: 'Yes, you were on it. You returned to your [life]boat number 5?'

Pitman: 'Yes.'

Aspinall: 'When you got back, were any people being put into it [the lifeboat]?'

Pitman: 'None at all.'

Aspinall: 'What happened then?'

Pitman: 'I simply stood in the [life]boat and said, "Come along, ladies," and helped them in – Mr Ismay helped to get them there.

Aspinall: 'How many ladies did you get in?'

Pitman: 'I do not know; between 30 and 40, I should imagine.'

Aspinall: 'Were there any children?'

Pitman: 'Yes, we had two [children].'

Aspinall: 'Could you tell whether these women were first, second or third class passengers that were getting into the [life]boat?'

Pitman: 'Most, I should say, would be first class.'

Aspinall: 'In addition to those women that you got into the [life]boat, did you take any male passengers in?'

Pitman: 'Yes, I should say about half a dozen or more.'

Aspinall: 'Why did you let the male passengers in?'

Pitman: 'Simply because there were no more women around – at least, there were two there, but they would not come.'

Aspinall: 'Did they give you any reason for refusing to come?'

Pitman: 'No.'

Aspinall: 'You say there were no other women around? Could you see whether there were other women in other parts of the boat deck? Did you notice at that time?'

Pitman: 'There were none in sight at that time – at least, not on the starboard deck.'

Aspinall: 'In view of the number that you had got into the [life]boat at this time, did you think that that was as many as this [life]boat would safely carry before she was lowered to the water?'

Pitman: 'No, I did not decide how many she should take.'

Aspinall: 'Who decided that?'

Pitman: 'Mr Murdoch, he came along just then.'

Aspinall: 'What did he say?'

Pitman: 'Well, I jumped out of the [life]boat then, ready to lower away, and he said, "You go in charge of this [life]boat, and also look after the others, and stand by to come along the after gangway when hailed."'

Aspinall: 'Did you go in charge of this [life]boat?'

Pitman: 'I did.'

Aspinall: 'There were 30 to 40 women you have told us, two children, about half a dozen male passengers, yourself, and how many of the crew?'

Pitman: 'Four.'

Commissioner: 'Did the four include yourself?'

Pitman: 'No, My Lord.'

Aspinall: 'Did you say something about Mr Murdoch saying he would hail you when he wanted you alongside the gangway?'

Pitman: 'Yes. He said, "Keep handy to come to the after gangway." Therefore, I understood he would hail us.'

Aspinall: 'You understood it?'

Pitman: 'Yes.'

Aspinall: 'Was the [life]boat properly lowered away?'

Pitman: 'It was.'

Aspinall: 'And you got put down to the water's edge?'

Pitman: 'Yes.'

Aspinall: 'On reaching the water what was done with that [life]boat?'

Pitman: 'We pulled away about 100 yards from the side of the ship.'

Aspinall: 'And then?'

Pitman: 'Lay on our oars.'

Aspinall: 'Did you take her in the direction of the gangway, in case Mr Murdoch might hail you and order you back?'

Pitman: 'Well, we dropped astern a little.'

Aspinall: 'That would be somewhere in the direction towards the gangway?'

Pitman: 'Yes.'

Commissioner: 'Just put your finger on the gangway you are talking about.

Pitman: [Pointed it out on the model.]

Aspinall: 'That is right aft?'

Pitman: 'Yes, he said the after gangway.'

Aspinall: 'Before you left the ship had you heard any order given about lowering the gangway or opening the gangway door?'

Pitman: 'No, that was the first I knew of it.'

Aspinall: 'After you were in the [life]boat and had rowed out this 100 yards somewhat astern did you notice whether the gangway door was open or not?'

Pitman: 'I do not think it was.'

Aspinall: 'You probably were looking in that direction?'

Pitman: 'Well, I was watching the ship the whole time.'

Aspinall: 'And you do not think it was opened?'

Pitman: 'I do not.'

Commissioner: 'How many gangways are there that side of the ship?'

Aspinall: 'I think two, My Lord, but I speak subject to correction.'

Commissioner: 'The gangway we have heard of yesterday was forward.'

Aspinall: 'Yes.'

Solicitor General: 'Yesterday we did hear of them at both ends.'

Aspinall: '12 or 13 [gangways] the builders tell me.'

Commissioner: 'On each side?'

Aspinall: 'No. I am told there are eight passenger gangways.'

Commissioner: 'Does that mean four on each side?'

Aspinall: 'Yes.'

Commissioner: 'But they are not on the same level.'

Aspinall: 'Two on D Deck amidships – one on E Deck forward and aft; that would make four on the one side and four on the other. If I might go back for one moment – I do not know whether it is important or not, but it might become important – did Mr Murdoch, in addition to telling you to keep handy to come back to the gangway, say anything more to you?'

Pitman: 'No; he only shook hands and said, "Goodbye, good luck"; that was all.'

Aspinall: 'When he said "Goodbye" to you in that way, did you think the situation was serious; did you think the ship was doomed then?'

Pitman: 'I did not, but I thought he must have thought so.'

Aspinall: 'Again, with regard to the time, how long do you think it was between the time of striking the [ice]berg and your [life]boat reaching the water. You have given me two estimates of time, 20 minutes, and 12.20 [am]. Could you help me on this matter?'

Pitman: 'Well, I should think it would be about 12.30 [am] when number 5 [life]boat reached the water.'

Aspinall: 'I do not know whether this will help you to see whether that is right. Was your [life]boat in the water about an hour before the *"Titanic"* went down?'

Pitman: 'I think it was longer than that.'

Aspinall: 'Much longer or a little longer?'

Pitman: 'It is hard to say.'

Aspinall: 'Now, I have got you in the [life]boat somewhere about 100 yards from the ship, you [are] watching the ship. Whilst you were watching the ship did you then begin to think she was in a condition in which it was probable she might be lost?'

Pitman: 'No, I did not give up hopes until I saw the last line of lights on the forecastle head disappear.'

Aspinall: 'When you reached the water and were in the [life]boat, did you see then that her head was getting deeper and deeper in the water?'

Pitman: 'Oh, yes, I watched the different lines of lights disappear.'

Aspinall: 'Did you see any other [life]boat on the water anywhere near you after your [life]boat had reached the water?'

Pitman: 'Are you alluding to one of our [life]boats?'

Aspinall: 'Yes, I mean one of the *"Titanic"* [life]boats?'

Pitman: 'Yes, [lifeboat] number 7 was quite close to me.'

Aspinall: 'Was [lifeboat] number 7, as far as you know, in the water before yours or after?'

Pitman: '[Lifeboat] number 7 was before; it was the first [life]boat launched on the starboard side.'

Aspinall: '[Lifeboat] number 7?'

Pitman: '[Lifeboat] number 7.'

Aspinall: 'And the second [life]boat was?'

Pitman: '[Lifeboat] number 5, and [lifeboat] number 3 next.'

Aspinall: 'How do you know [lifeboat] number 3 came next? Did you see it?'

Pitman: 'I saw it coming down; I saw it being lowered.'

Aspinall: 'Did you notice any other [life]boats on that side being lowered?'

Pitman: 'I did not.'

Aspinall: 'You speak of [lifeboats] 7, 5, and 3?'

Pitman: '[Lifeboats] 7, 5, and 3, yes.'

Aspinall: 'In that order? The solicitor general points out that [Archie] Jewell refers to this that [lifeboat] number 7 was the first [life]boat on the starboard side.'

Solicitor General: 'At page 19, question 147, [Archie] Jewell says he was in the [life]boat, and it was the first to go on the starboard side.'

Pitman: 'That is right.'

Aspinall: 'Tell me with regard to the equipment of the [life]boat you were in, do you know whether it had a lamp or not?'

Pitman: 'Mine had not.'

Aspinall: 'Did you look for it?'

Pitman: 'I did.'

Aspinall: 'And would you, as an officer, know what was the right place to look for the lamp?'

Pitman: 'Exactly.'

Aspinall: 'Was there any compass in your [life]boat?'

Pitman: 'No.'

Aspinall: 'Did you look for it?'

Pitman: 'Well, I did not at the time, because it would be absolutely useless to me.'

Aspinall: 'But how do you know there was no compass? You say you did not look at the time. Did you look at some later time?'

Pitman: 'Yes, after the [life]boats were on the *"Carpathia".*'

Aspinall: 'Was there any water in your [life]boat?'

Pitman: 'Yes.'

Aspinall: 'In what? Breakers?'

Pitman: 'In two breakers.'

Aspinall: 'Two breakers?'

Pitman: 'Yes.'

Aspinall: 'Were there any biscuits?'

Pitman: 'Yes.'

Aspinall: 'In what?'

Pitman: 'A tank in the stern of the [life]boat.'

Aspinall: 'Whilst you were in the [life]boat and before the ship sank, did you see any light or lights which you took to be the light or lights of another steamer?'

Pitman: 'I saw a white light which I took to be the stern light of a sailing ship.'

Aspinall: 'How far away did you judge it to be?'

Pitman: 'I thought it was about five miles.'

Aspinall: 'That would be a good distance to see a stern light, would it not?'

Pitman: 'Yes, it may have been less.'

Aspinall: 'Was it a good night for seeing a light; for seeing a good stern light?'

Pitman: 'An excellent night.'

Aspinall: 'They would be visible at a long distance?'

Pitman: 'Yes.'

Aspinall: 'Whilst you were in the [life]boat did you notice the *"Titanic"* sending up rockets?'

Pitman: 'Yes, she did.'

Aspinall: 'We have heard this in detail. Was there good discipline and order maintained in your [life]boat?'

Pitman: 'Well, that is not for me to say; it is for other people to say that.'

Commissioner: 'No, but you are asked your opinion?'

Pitman: 'As regards the passengers, yes, and the crew.'

Aspinall: 'I am not suggesting you did not behave well; I am only asking the question for the information of the court. It is a general question asked with regard to all the [life]boats. I am not suggesting for one moment that there was anything wrong. You behaved well, I have no doubt?'

Pitman: 'I do not know about myself; it is not for me to say that.'

Aspinall: 'You say the passengers and the crew behaved well?'

Pitman: 'They did.'

Aspinall: 'Well, that exhausts it. Now you saw the vessel go down?'

Pitman: 'Yes.'

Aspinall: 'What did she do when she went down; you were an officer, perhaps you can tell us. Inquires have been made of others. How did she sink? She sank by the head, we know that?'

Pitman: 'Yes.'

Commissioner: 'Just describe it in your own way.'

Aspinall: 'May I hand him the profile.'

Commissioner: 'Yes.'

Commissioner: [To Pitman] 'Describe it in your own way.'

[The profile was handed to Pitman.]

Pitman: 'That is the position I saw her in when we left. She gradually disappeared like that; she went right on end like that and went down that way'

Pitman: [Demonstrating.]

Commissioner: 'Did her afterpart ever right itself?'

Pitman: 'I should not think so; I did not see it.'

Commissioner: 'Before she finally disappeared?'

Pitman: 'No.'

Commissioner: 'Could you have seen it if it had happened?'

Pitman: 'I think so; I was only barely 100 yards away.'

Commissioner: 'Were you keeping your eyes upon her?'

Pitman: 'I was.'

Commissioner: 'You know this is suggested – supposing that is the head of the ship and going down in this way with the afterpart coming up in that way; a number of witnesses have said that before she finally foundered, plunged into the sea, the afterpart righted itself like that and then she went down. The question is whether you think that is true that she broke in two in that way bringing her afterpart level with the water again and then went down in that way. Did she crack in the middle?'

Pitman: 'I do not think so. If the afterpart had broken off it would have remained afloat.'

Commissioner: 'Not broken off, but cracked in that way?'

Pitman: 'No.'

Commissioner: 'At all events, the point is this: Did you see the afterend of the ship – you saw it up in the air – right itself and come flush with the water again?'

Pitman: 'It did not.'

Commissioner: 'And you say you looked, and if it had happened you would have seen it?'

Pitman: 'Certainly.'

Aspinall: 'While you were in the water, before the *"Titanic"* sank, did you hear any hail either from Mr Murdoch or the captain [Smith] or from anybody else to come back near the gangway?'

Pitman: 'No.'

Aspinall: 'Did you hear anybody on the *"Titanic"* using a megaphone?'

Pitman: 'I did not.'

Aspinall: 'Did you transfer any of your passengers to any other [life]boat?'

Pitman: 'Yes, I transferred four, I think it was.'

Aspinall: 'Into what [life]boat?'

Pitman: 'I am not quite certain of the number, but I think it was [lifeboat] number 7.'

Aspinall: 'Why did you transfer those passengers?'

Pitman: 'Because they had a less number of passengers in that [life]boat than I had.'

Aspinall: 'As the "*Titanic*" sank and immediately after did you hear any screams?'

Pitman: 'Immediately after she sank?'

Aspinall: 'Yes?'

Pitman: 'Yes.'

Aspinall: 'Were you able to go in the direction of the screams and render any assistance?'

Pitman: 'I did not go.'

Aspinall: 'But do you think you could have gone? I am not suggesting anything; I only want to get the facts from you. Do you think it would have been safe or reasonable to go?'

Pitman: 'I do not.'

Aspinall: 'What is your reason?'

Pitman: 'Well, there was such a mass of people in the water we should have been swamped.'

Aspinall: 'In your view you had a sufficient number of people on your [life]boat. Is that so?'

Pitman: 'No, but I had too many in the [life]boat to go back to the wreck.'

Aspinall: 'And I think you remained on, the men more or less lying on their oars till daylight, and then you were picked up by the "*Carpathia*"?'

Pitman: 'Yes; we lay at rest the remainder of the night.'

Commissioner: 'Before you go into that there are two questions I want to put.'

Commissioner: [To Pitman] 'Whereabouts were you when the "*Titanic*" sank?'

Pitman: 'About 200 yards away.'

Commissioner: 'On what side?'

Pitman: 'On the starboard quarter.'

Commissioner: 'Would that be about abreast of the mainmast?'

Pitman: 'About that, My Lord.'

Aspinall: 'When you gave evidence in America you said this: I want to know if you say it is accurate. You were asked: "Can you fix the exact moment of time when the '*Titanic*' disappeared?"'

Pitman: 'Two-twenty exactly, ship's time. I took my watch out at the time she disappeared, and I said, "It is 2.20 [am]," and the passengers around me heard it.'

Aspinall: 'Do you remember giving that evidence?'

Pitman: 'That is true, yes.'

Aspinall: 'That is correct?'

Pitman: 'Yes.'

Aspinall: 'I want you to give me the benefit of your views on this matter. One of the questions which will probably be asked is this: "Had the *'Titanic'* the means of throwing searchlights around her. If so, did she make use of them to discover ice? Should searchlights have been provided and used?" In view of your experience of these waters and also as an officer and a sailor, what is your view as to the utility of searchlights when you are in the ice region?'

Pitman: 'They might be of some assistance.'

Commissioner: 'Have you ever seen them used?'

Pitman: 'No, My Lord.'

Aspinall: 'Have you ever considered the matter before I asked you the question?'

Pitman: 'No, I have never considered it before.'

Commissioner: 'Do you know whether searchlights are used upon any Atlantic liners for the purpose of ascertaining whether there is ice?'

Pitman: 'I never heard of it, My Lord.'

Thomas Scanlan MP: 'During your watches from 6 to 8 [pm] and previously from 12 till 4 [pm] were you in charge of the bridge?'

Pitman: 'No.'

Scanlan: 'Who was the officer with you?'

Pitman: 'The chief [Wilde], 2 to 6 [pm] in the afternoon.'

Scanlan: 'From 12 to 4 [pm]?'

Pitman: 'The first [Murdoch] and chief [Wilde].'

Scanlan: 'And from 6 to 8 [pm]?'

Pitman: 'The second [Lightoller].'

Scanlan: 'Was the whole of the knowledge that you had of icebergs obtained from the chart?'

Commissioner: 'From the chart room.'

Scanlan: 'From the chart in the chart room?'

Pitman: 'No, from the Marconigrams.'

Scanlan: 'Were any Marconigrams handed to you from 12 to 4 [pm]?'

Pitman: 'Not to me.'

Scanlan: 'Had you seen any Marconigram that reached the ship with reference to ice from 12 to 4 [pm] on Sunday?'

Pitman: 'I saw two that reached the ship that day. I have no idea what time they arrived.'

Scanlan: 'You saw two that day?'

Pitman: 'Yes.'

Scanlan: 'Do you know whether they were Marconigrams that had come the previous day, on the Saturday?'

Pitman: 'No, they came on the Sunday.'

Scanlan: 'Are you quite sure of that?'

Pitman: 'Yes.'

Scanlan: 'Did you read them?'

Pitman: 'Yes, I read one – yes.'

Scanlan: 'You read one, but you did not read the other?'

Pitman: 'No, not to remember what was on it.'

Scanlan: 'With respect to the one that you did read, can you tell us what was on it?'

Pitman: 'No, no more than it stated, "Ice in longitude 49 to 51 W [west]."'

Scanlan: 'Where did it come from?'

Pitman: 'I have no idea.'

Scanlan: 'Had it come on the Sunday?'

Pitman: 'It must have done. Had it come on the Saturday I should have seen it before.'

Scanlan: 'When ice is reported to you, is it the duty of someone on the bridge or in the chart room to indicate on the chart kept in the chart room the location of the ice?'

Pitman: 'I do not know about its being duty; we often do it; in fact, we usually do it.'

Scanlan: 'I see you gave evidence on this matter in America. You said in answer to Senator Smith:
[Senator Smith] "You stated a few minutes ago that the second officer [Lightoller], I believe, reported ice on the Saturday night?"
(Mr Pitman) "No, I said the fourth officer [Boxhall]." (Senator Smith) "Mr Lowe?"
(Mr Pitman) "Mr Boxhall." Did you give this evidence:
(Senator Smith) "You said Mr Boxhall reported ice Saturday night, and that it was marked on the chart with a cross"?'

Pitman: 'That is a mistake. It is Sunday night.'

Scanlan: 'It is a mistake?'

Pitman: 'Yes, it was Sunday night.'

Scanlan: 'Where was it marked?'

Pitman: 'On the North Atlantic Track Chart.'

Scanlan: 'I know it was marked on the chart, but where was it marked with reference to the course you were steering?'

Pitman: 'Some miles north of it.'

Scanlan: 'It was marked some miles north of the course you were steering. You were also questioned as to whether you had been made aware on the Sunday of any message which the *"Titanic"* had received from the *"Californian"* about ice. You were asked: "Did you learn from Mr Lightoller that the *'Californian'* had warned the *'Titanic'* that she was in the vicinity of icebergs?" (Mr Pitman) "I did not, sir. We had no conversation whatever." (Senator Smith) "Did you hear anything about a wireless from the *'Californian'* on the direction of icebergs?" (Mr Pitman.) "I did not, sir." Is that the true state of the facts?'

Pitman: 'That is so, yes; I did not know anything about the *"Californian"* till the Monday morning.'

Scanlan: 'That is, although you were an officer on the bridge from 6 to 8 [pm] you know nothing of any wireless having come from the *"Californian"*?'

Pitman: 'None came from 6 to 8 pm.'

Scanlan: 'Or from any other ship?'

Solicitor General: 'Do you mean the *"Californian"*, Mr Scanlan?'

Scanlan: 'Yes.'

Scanlan: [To Pitman] 'I think they did get a message about 6 [pm]?'

Pitman: 'No messages arrived between those hours.'

Scanlan: 'Had any message arrived that day that you knew of?'

Pitman: 'The two Marconigrams I mentioned before arrived that day. That is all I know of.'

Commissioner: 'You are talking about the Sunday?'

Scanlan: 'Yes.'

Commissioner: 'He has told us he saw two Marconigrams on the Sunday which had not arrived on the Saturday, but that he does not know what ships they came from.'

Scanlan: [To Pitman] 'Was any mark put on the chart on the Sunday with reference to any messages you received on the Sunday?'

Pitman: 'Yes, as far as I can remember, one was put on the chart between 4 and 6 [pm].'

Day Fourteen, 23 May (re-called)

Sir Robert Finlay: 'You were the third officer as we know?'

Pitman: 'Yes.'

Sir Robert: 'You were on duty, we have heard, from 12 to 4 [pm] in the afternoon, and then from 6 to 8 [pm] on this Sunday?'

Pitman: 'That is correct.'

Sir Robert: 'Did you ever hear anything about the message from the *"Amerika"* stated to have been received about 2 pm for transmission to Cape Race, or about the *"Mesaba"* message?'

Pitman: 'Nothing whatever.'

Solicitor General: 'I will put the same question to you. How many messages about ice on this 14th [April 1912] you know of?'

Pitman: 'I can only recollect one.'

Solicitor General: 'We have been told that messages are posted in the chart room. Did you see any message posted in the chart room?'

Pitman: 'Yes, only one; that is the *"Caronia"* message.'

Solicitor General: 'You are quite clear about that?'

Pitman: 'Well, there were a few other messages posted there, but they related to the time we left Southampton.'

Solicitor General: ' We are talking about ice?'

Pitman: 'Yes, there was ice on those reports.'

Solicitor General: 'I refer to what you said on page 346. I am going to read you five or six questions and answers beginning at 14921. Mr Asquith asked you: [Asquith] "Was it within your knowledge that the ship would probably meet with ice that evening?"
(A) [Pitman] "We knew that we should be in the longitude of ice."
(Q) [Asquith] "Who told you that?" – (A) [Pitman] "I saw it in a Marconigram."
(Q) [Asquith] "Do you know from which ship that Marconigram had come?"
(A) [Pitman] "I have no idea."
(Q) [Asquith] "Did you particularly concern yourself with that matter?"
(A) [Pitman] "No. I simply looked at them, and saw that there was no ice reported on the track."
(Q) [Asquith] "Now listen to this: Did you see one Marconigram or two Marconigrams?"
(A) [Pitman] "Two, I think" Is that right?'

Pitman: 'No, I am not absolutely certain about that.'

Solicitor General: 'But you were absolutely certain three minutes ago that there was only one, were not you?'

Pitman: 'Yes, I can only recall one.'

Solicitor General: 'Just get your memory, now, to serve you as well as it can. I am sure you will. Just tell us frankly. Do you remember posted in the chart room one, or more than one, Marconigram?'

Pitman: 'I am not certain.'

Solicitor General: 'I do not want to treat you other than fairly, but just listen to the next answer. You were asked, [Interviewer] "Were these posted – those are the two you have just spoken of – in some part of the ship?"
(A) [Pitman] "Yes, in the chart room."
(Q) [Interviewer] "And you read them, did you?"

(A) [Pitman] "Yes." Is it not clear to you that when you gave evidence on the 22nd May [1912] your then impression was that there were two separate messages about ice posted in the chart room.'

Sir Robert: 'It is quite clear he says, yes. I think.'

Pitman: 'It may be, but I am not clear on that.'

Solicitor General: 'When did you cease to be clear?'

Pitman: 'I have forgotten that I said there were two there.'

Commissioner: 'I think the right question would be, when was he first clear, if ever.'

Solicitor General: 'You will see it was not the learned counsel that suggested two to you; he suggested that you had seen one, and you corrected him and said, "No, two."'

Sir Robert: 'Forgive me; what the learned counsel said was: "Did you see one Marconigram or two Marconigrams?" and the witness [Pitman] says: "Two, I think."'

Solicitor General: 'I think I am quite right. Two questions further up he [Pitman] was asked:
[Interviewer] "Do you know from which ship that Marconigram had come?"
and his [Pitman's] answer was:
[Pitman] "I have no idea."
(Q) [Interviewer] "Did you particularly concern yourself with that – matter?" Up to that time Mr Aspinall has not known anything of more than that one. The witness
[Pitman] says "No, I simply looked at them."
And then he [Pitman] is asked:
[Interviewer] "Did you see one Marconigram or two Marconigrams?" and he [Pitman] says:
[Pitman] "Two, I think"?'

Pitman: 'I think it quite possible there were two there, but one related to the oil tank steamer.'

Sir Robert: 'Let us keep to the same point. Did not you understand me just now to be asking about messages about ice?'

Pitman: 'Yes.'

Sir Robert: 'And did not your answers refer to messages about ice?'

Pitman: 'No, I think you asked me about Marconigrams on the notice board.'

Sir Robert: 'Is that the explanation?'

Pitman: 'Yes.'

Sir Robert: 'Now I will put it beyond the possibility of doubt. As far as your knowledge goes, Mr Pitman, had you ever seen or heard of more than one Marconigram about ice on the 14th April [1912]?'

Pitman: 'One only.'

Solicitor General: 'Now, My Lord, May I refer you to page 349.'

Solicitor General: [To the Pitman] 'I want you to hear your answers and just consider. You were asked by Mr Scanlan at Question 15107: [Scanlan] "Was the whole of the knowledge that you had of icebergs obtained from the chart?"
[Commissioner] 'From the chart room.' [Scanlan] 'From the chart in the chart room?' –
(A) [Pitman] "No, from the Marconigrams."
That is more than one, is it not? – Then the next question is: [Scanlan] "Were any Marconigrams handed to you from 12 to 4 [pm]?"
(A) [Pitman] "Not to me."
(Q) [Scanlan] "Had you seen any Marconigram that reached the ship with reference to ice from 12 to 4 [pm] on Sunday?"
(A) [Pitman] "I saw two that reached the ship that day."
Now, is that right?'

Pitman: 'That is correct. They did not relate to ice.'

Solicitor General: 'I will read the question again. The question was: [Interviewer] "Had you seen any Marconigram that reached the ship with reference to ice from 12 to 4 [pm] on Sunday?" That was the question. Did you hear it?'

Pitman: 'Yes.'

Solicitor General: 'That is with reference to ice, do you see?'

Pitman: 'They did not relate to ice.'

Solicitor General: But your answer is: [Pitman] "I saw two that reached the ship that day. I have no idea what time they arrived." Then I ought to read on, I think:
(Q) [Scanlan] "You saw two that day?"
(A) [Pitman] "Yes."
(Q) [Scanlan] "Do you know whether they were Marconigrams that had come the previous day, on the Saturday?"
(A) [Pitman] "No, they came on the Sunday."
(Q) [Scanlan] "Are you quite sure of that?"
(A) [Pitman] "Yes."
(Q) [Scanlan] "Did you read them?"
(A) [Pitman] "Yes, I read one – yes."
(Q) [Scanlan] "You read one, but you did not read the other?" (A) [Pitman] "No, not to remember what was on it."
Now, do you mean to represent that when you were asked those questions you thought Mr Scanlan was asking about the tank steamer?'

Pitman: 'No, I do not know what he was relating to.'

Solicitor General: 'The next question was:

[Scanlan] "With respect to the one that you did read, can you tell us what was on it?"
(A) [Pitman] "No, no more than it stated. 'Ice in longitude 40 to 51 W [west].'"
(Q) [Scanlan] "Where did it come from?"
(A) [Pitman] "I have no idea."
Is that the best you can tell us about that matter?'

Pitman: 'That is it.'

Commissioner: 'That looks as if he had only read one telegram.'

Solicitor General: 'Yes, but that is a plain question.'

Sir Robert: [To the Pitman] 'You went off duty at 8 [pm]?'

Pitman: 'Yes.'

Sir Robert: 'You were third officer, as we know?'

Pitman: 'Yes.'

Sir Robert: 'You were on duty, we have heard, from 12 to 4 [pm] in the afternoon, and then from 6 to 8 [pm] on this Sunday?'

Pitman: 'That is correct.'

Fourth Officer Joseph Groves Boxhall

Joseph Boxhall was questioned in the United States on 22 April 1912, and re-called on 29 April 1912; and at the British inquiry on days thirteen and fourteen as follows:

Day Thirteen, 22 May 1912

Raymond Asquith: 'Were you the fourth officer of the *Titanic* at the time of the accident?'

Boxhall: 'I was.'

Asquith: 'What certificate do you hold?'

Boxhall: 'Extra master.'

Asquith: 'You have held that, I think, for about 4 or 5 years?'

Boxhall: 'September, 1907.'

Asquith: 'Have you been employed for five years by the White Star company?

Boxhall: 'Five years next November.'

Asquith: 'And has most of your experience been in the Atlantic?'

Boxhall: 'Most of it, yes.'

Asquith: 'On Sunday the 14th April [1912], were you on watch from 4 to 6 in the evening?'

Boxhall: 'I was.'

Asquith: 'Who else was on the bridge at that time?'

Boxhall: 'Mr Wilde, the chief officer, and Mr Moody, the sixth officer.'

Asquith: 'Can you say what the course of the ship was when you came on watch at 4 o'clock?'

Boxhall: 'No; I have forgotten the course.'

Asquith: 'The course would be marked, I suppose, on a course board?'

Boxhall: 'Oh, yes, the course was there.'

Asquith: 'You cannot say what it was?'

Boxhall: 'No, I cannot remember.'

Asquith: 'Between 4 and 6 [pm], while you were on watch do you remember the course being altered?'

Boxhall: 'The course was altered at 5.50 [pm].'

Asquith: 'Do you remember what it was altered to?'

Boxhall: 'I do not remember the compass course, but I remember the true course was S [south] 86 W [west].'

Asquith: 'I think you worked that out yourself?'

Boxhall: 'Yes, I had stellar observations afterwards.'

Asquith: 'At the time when you came on watch at 4 o'clock had you heard anything about ice being in the neighbourhood?'

Boxhall: 'Yes, I had seen reports of ice and put them on the chart.'

Asquith: 'Reports which had been received earlier on the same day, do you mean?'

Boxhall: 'No, I cannot say from my own recollection that they were received on Sunday, but subsequently I have heard that some of them, or one of them that I put on the chart, was received on the Sunday, and that I put it on between 4 and 6. All the ice marked on the captain's [Smith's] chart I put down myself.'

Asquith: 'Do you remember what these messages indicated with regard to ice?'

Boxhall: 'Yes, it indicated the region of the ice.'

Asquith: 'Do you remember what region it indicated; did it convey to your mind that you at 4 o'clock were somewhere in the neighbourhood of ice or not?'

Boxhall: 'It conveyed to my mind that the ship would shortly be in the region of the ice.'

Asquith: 'Did you make up your mind about what time that would happen?'

Boxhall: 'No, I did not.'

Asquith: 'Was the position of the ice marked upon the chart when you came up at 4 o'clock, I mean the position in which it might be expected?'

Boxhall: 'Some of the positions were on the chart that I had put on previously.'

Commissioner: 'But I do not know when it was that you marked the chart?'

Boxhall: 'I do not remember myself, My Lord.'

Commissioner: 'But it was on the Sunday?'

Boxhall: 'Well, subsequently I have heard it was on the Sunday, between 4 and 6 [pm], that I put some of the positions on.'

Solicitor General: 'I fancy Mr Boxhall has not been very well. I know your Lordship will allow him to sit down.'

Commissioner: 'Certainly, sit down, if you wish?'

Boxhall: 'I am quite right, My Lord, thank you.'

Asquith: 'You cannot recollect when it was you marked the position of the ice on the chart?'

Boxhall: 'No. Some of the positions were from the "*La Touraine*". Well, that must have been a couple of days previously, I think.'

Commissioner: 'And had they reference to ice?'

Boxhall: 'Yes, ice and derelicts.'

Asquith: 'I understand your recollection is that during the period between 4 and 6 [pm] you did not make any additional mark on the chart?'

Boxhall: 'To my recollection, I did not, but others say that I did.'

Asquith: 'Do you remember whether, when you went off watch at 6 [pm] you noticed any marks on the chart with regard to ice which you had not noticed before?'

Boxhall: 'No, I cannot say I noticed any.'

Asquith: 'Then you went off watch at 6 o'clock, and came on again, I think, at 8 o'clock the same evening?'

Boxhall: 'Yes.'

Asquith: 'When you came up at 8 o'clock, was Mr Lightoller on the bridge in charge?'

Boxhall: 'Mr Lightoller was there.'

Asquith: 'Did you look at the chart then. Do you remember whether there was anything new about the ice marked on the chart at 8 o'clock?'

Boxhall: 'I did not look at the chart when I came on at 8 o'clock.'

Asquith: 'Your watch was from 8 to 12 [pm], was it not?'

Boxhall: 'Yes.'

Asquith: 'Do you remember during that period whether any messages were received with regard to ice, upon the bridge?'

Boxhall: 'No, I do not recollect any.'

Asquith: 'You have no recollection of a message from the "*Californian*" or the "*Antillian*" being brought to the bridge?'

Boxhall: 'No.'

Asquith: 'Were you on the bridge, looking out, most of that time, or were you somewhere else making calculations?'

Boxhall: 'I was inside the chart room working up stellar observations from 8 o'clock.'

Asquith: 'We have heard it was a fine, clear cold night. Is that your recollection?'

Boxhall: 'Yes, it was perfectly clear.'

Asquith: 'Did you see anything in the nature of haze?'

Boxhall: 'No, none whatever.'

Asquith: 'Neither at 8 o'clock nor at any time during the night?'

Boxhall: 'Whenever I was on the deck or at the compass I never saw any haze whatever.'

Asquith: 'And did you think, when you came up at 8 [pm] that the ship was nearing the neighbourhood of ice?'

Boxhall: 'It did not enter my mind.'

Asquith: 'Was the first intimation that there was ice about the striking of the three bells, so far as you were concerned?'

Boxhall: 'No, when we struck the [ice]berg; that was the first.'

Asquith: 'Do you mean you felt the shock before you heard the bells?'

Boxhall: 'No, I heard the bells first.'

Asquith: 'Where were you at that time?'

Boxhall: 'Just coming out of the officers' quarters.'

Asquith: 'How soon after you heard the bells did you feel the shock?'

Boxhall: 'Only a moment or two after that.'

Asquith: 'Did you hear an order given by the first officer [Murdoch]?'

Boxhall: 'I heard the first officer [Murdoch] give the order, "Hard-a-starboard," and I heard the engine room telegraph bells ringing.'

Asquith: 'Was that before you felt the shock, or afterwards?'

Boxhall: 'Just a moment before.'

Commissioner: 'Let us be clear about that. The order, "Hard-a-starboard," came between the sound of the bells and the collision?'

Boxhall: 'The impact, yes.'

Asquith: 'Did you go on to the bridge immediately after the impact?'

Boxhall: 'I was almost on the bridge when she struck.'

Asquith: 'Did you notice what the telegraphs indicated with regard to the engines?'

Boxhall: '"Full speed astern," both.'

Asquith: 'Was that immediately after the impact?'

Boxhall: 'Yes.'

Asquith: 'Did you see anything done with regard to the watertight doors?'

Boxhall: 'I saw Mr Murdoch closing them then, pulling the lever.'

Asquith: 'And did the captain [Smith] then come out on to the bridge?'

Boxhall: 'The captain [Smith] was alongside of me when I turned round.'

Asquith: 'Did you hear him [Smith] say something to the first officer [Murdoch]?'

Boxhall: 'Yes, he [Smith] asked him [Murdoch] what we had struck.'

Asquith: 'What conversation took place between them [Smith and Murdoch]?'

Boxhall: 'The first officer [Murdoch] said, "An iceberg, sir. I hard-a-starboarded and reversed the engines, and I was going to hard-a-port round it but she was too close. I could not do any more. I have closed the watertight doors." The commander [Smith] asked him if he had rung the warning bell, and he said "Yes."'

Asquith: 'Did the captain [Smith] and the first officer [Murdoch] go to the starboard side of the bridge to see if they could see the iceberg?'

Boxhall: 'Yes.'

Asquith: 'Did you see it yourself?'

Boxhall: 'I was not too sure of seeing it. I had just come out of the light, and my eyes were not accustomed to the darkness.'

Asquith: 'What did you do next – did you leave the deck?'

Boxhall: 'Yes, I went down forward, down into the third class accommodation, right forward on to the lowest deck of all with passenger accommodation, and walked along these looking for damage.'

Asquith: 'That would be F Deck, would it not?'

Boxhall: 'Yes, F Deck. I walked along there for a little distance just about where I thought she had struck.'

Asquith: 'Did you find any signs of damage?'

Boxhall: 'No, I did not.'

Commissioner: 'What deck is it?'

Asquith: 'F Deck, he says.'

Commissioner: 'You say it is F Deck?'

Boxhall: 'I am not quite sure, My Lord, but it was the lowest deck I could get to without going into the cargo space.'

Asquith: 'The lowest deck on which there is passenger accommodation, he said.'

Commissioner: 'Is not that G [Deck]?'

Asquith: 'Yes, My Lord, I think it must be G [Deck].'

Commissioner: 'It is pointed out that he could not get along G Deck, because there is no door in the bulkhead, and therefore it cannot have been G Deck.'

Asquith: [To Boxhall] 'How did you get down to the lowest of these decks which you went to?'

Boxhall: 'Through a staircase under the port side of the forecastle head which takes me down into D Deck, and then walked along aft along D Deck to just underneath the bridge, and down the staircase there on the port side, and then I am down on E Deck near E Deck doors, the working alleyway; and then you cross over to the starboard side of E Deck and go down another accommodation staircase on to F Deck. I am not sure whether I went lower. Anyhow, I went as low as I could possibly get.'

Commissioner: [To Boxhall] 'Just come round here?'

Boxhall: 'Yes, My Lord.'

Boxhall: [Explains the plan to the commissioner.]

Commissioner: 'He appears to have got to F Deck. His first statement was right.'

Asquith: [To Boxhall] 'Did you then go up again through the other decks as far as C Deck?'

Boxhall: 'I came up the same way as I went down.'

Asquith: 'Without noticing any damage?'

Boxhall: 'I did not see any damage whatever.'

Asquith: 'When you got to C Deck did you see some ice there on the deck?'

Boxhall: 'Yes, I took a piece of ice out of a man's hand, a small piece about as large as a small basin, I suppose; very small, anyhow; about that size [Describing]. He was going down again to the passenger accommodation, and I took it from him and walked across the deck to see where he got it. I found just a little ice in the well deck covering a space of about three or four feet from the bulwarks right along the well deck, small stuff.'

Asquith: 'Did you then go and report to the captain [Smith]?'

Boxhall: 'I went on to the bridge and reported to the captain [Smith] and first officer [Murdoch] that I had seen no damage whatever.'

Asquith: 'Did the captain [Smith] then tell you to find the carpenter?'

Boxhall: 'Yes, I think we stayed on the bridge just for a moment or two, probably a couple of minutes, and then he told me to find the carpenter and tell him to sound the ship forward.'

Asquith: 'Did you find the carpenter?'

Boxhall: 'I met the carpenter. I think it would be on the ladder leading from the bridge down to A Deck, and he wanted to know where the captain [Smith] was. I told him he was on the bridge.'

Asquith: 'Did the carpenter tell you anything about there being water?'

Boxhall: 'Yes, he did; he said the ship was making water fast, and he passed it on to the bridge.'

Asquith: 'What did you do?'

Boxhall: 'I continued with the intention of finding out where the water was coming in, and I met one of the mail clerks, a man of the name of Smith.'

Asquith: 'Did he say something?'

Boxhall: 'He also asked for the captain [Smith], and said the mail hold was filling. I told him where he could find the captain and I went down to the mail room. I went down the same way as I did when I visited the third class accommodation previously. I went down as far as E Deck and went to the starboard alleyway on E Deck and the watertight door stopped me getting through.'

Asquith: 'The watertight door on E Deck was closed?'

Boxhall: 'Yes. Then I crossed over and went into the working alleyway and so into the mail room.'

Asquith: 'What did you find in the mail room?'

Boxhall: 'I went down in the mail room and found the water was within a couple of feet of G Deck, the deck I was standing on.'

Asquith: 'The mail room is between the Orlop deck and G Deck?'

Boxhall: 'Yes, that is the mail hold.'

Asquith: 'Was the water rising or stationary?'

Boxhall: 'It was rising rapidly up the ladder and I could hear it rushing in.'

Asquith: 'Did you go back and report that to the captain [Smith] on the bridge?'

Boxhall: 'I stayed there just for a minute or two and had a look. I saw mail-bags floating around on deck. I saw it was no use trying to get them out so I went back again to the bridge. I met the second steward, Mr [George Charles] Dodd [1867–1912], on my way to the bridge – as a matter of fact in the saloon companion way – and he asked me about sending men down below for those mails. I said "You had better wait till I go to the bridge and find what we can do." I went to the bridge and reported to the captain [Smith].'

Asquith: 'We have been told that at some time you called the other officers; both Mr Lightoller and Mr Pitman said you called them?'

Boxhall: 'I did. That was after I reported to the captain about the mail room.'

Asquith: 'Could you form any opinion as to how long that was after the impact?'

Boxhall: 'No, but as near as I could judge; I have tried to place the time for it, and the nearest I can get to it is approximately 20 minutes to half an hour.'

Asquith: 'I think those are the times which are given by Mr Pitman and Mr Lightoller. After calling those officers did you go on to the bridge again?'

Boxhall: 'Yes, I think I went towards the bridge, I am not sure whether it was then that I heard the order given to clear the [life] boats or unlace the covers. I might have been on the bridge for a few minutes and then heard this order given.'

Asquith: 'Had you a [life]boat station of your own; did you know what it was?'

Boxhall: 'I did not know what it was.'

Asquith: 'We have been told it is customary for the third [Pitman] and fourth [Boxhall] officers to be assigned to the emergency [life] boats?'

Boxhall: 'Yes, it is for emergency purposes.'

Asquith: 'The third officer [Pitman] was assigned to [lifeboat] number 1. Were you assigned to [lifeboat] number 2?'

Boxhall: 'For emergency purposes I was assigned to [lifeboat] number 1 as a matter of fact, the starboard [life]boat.'

Asquith: 'When the order was given to clear the [life]boats what did you do; did you go to any particular [life]boat?'

Boxhall: 'No, I went right along the line of [life]boats and I saw the men starting, the watch on deck, our watch.'

Asquith: 'Which side of the ship?'

Boxhall: 'The port side, I went along the port side, and afterwards I was down the starboard side as well but for how long I cannot remember. I was unlacing covers on the port side myself and I saw a lot of men come along – the watch I presume. They started to screw some out on the afterpart of the port side; I was just going along there and seeing all the men were well established with their work, well under way with it, and I heard someone report a light, a light ahead. I went on the bridge and had a look to see what the light was.'

Asquith: 'Someone reported a light ahead?'

Boxhall: 'Yes; I do not know who reported it. There were quite a lot of men on the bridge at the time.'

Asquith: 'Did you see the light?'

Boxhall: 'Yes, I saw a light.'

Asquith: 'What sort of light was it?'

Boxhall: 'It was two masthead lights of a steamer. But before I saw this light I went to the chart room and worked out the ship's position.'

Asquith: 'Is that the position we have been given already – 41 deg. [degrees] 46 min. N [north], 50 deg [degrees], 14 min. W [west]?'

Boxhall: 'That is right, but after seeing the men continuing with their work I saw all the officers were out, and I went into the chart room to work out its position.'

Asquith: 'Was it after that you saw this light?'

Boxhall: 'It was after that, yes, because I must have been to the Marconi office with the position after I saw the light.'

Asquith: 'You took it to the Marconi office in order that it might be sent by the wireless operator?'

Boxhall: 'I submitted the position to the captain [Smith] first, and he told me to take it to the Marconi room.'

Asquith: 'And then you saw this light which you say looked like a masthead light?'

Boxhall: 'Yes, it was two masthead lights of a steamer.'

Asquith: 'Could you see it distinctly with the naked eye?'

Boxhall: 'No, I could see the light with the naked eye, but I could not define what it was, but by the aid of a pair of glasses I found it was the two masthead lights of a vessel, probably about half a point on the port bow, and in the position she would be showing her red if it were visible, but she was too far off then.'

Asquith: 'Could you see how far off she was?'

Boxhall: 'No, I could not see, but I had sent in the meantime for some rockets, and told the Captain [Smith] I had sent for some rockets, and told him I would send them off, and told him when I saw this light. He [Smith] said, "Yes, carry on with it." I was sending rockets off and watching this steamer. Between the time of sending the rockets off and watching the steamer approach us I was making myself generally useful round the port side of the deck.'

Asquith: 'How many rockets did you send up about?'

Boxhall: 'I could not say, between half a dozen and a dozen, I should say, as near as I could tell.'

Asquith: 'What sort of rockets were they?'

Boxhall: 'The socket distress signal.'

Asquith: 'Can you describe what the effect of those rockets is in the sky; what do they do?'

Boxhall: 'You see a luminous tail behind them and then they explode in the air and burst into stars.'

Asquith: 'Did you send them up at intervals one at a time?'

Boxhall: 'One at a time, yes.'

Asquith: 'At about what kind of intervals?'

Boxhall: 'Well, probably five minutes; I did not take any times.'

Asquith: 'Did you watch the lights of this steamer while you were sending the rockets up?'

Boxhall: 'Yes.'

Asquith: 'Did they seem to be stationary?'

Boxhall: 'I was paying most of my attention to this steamer then, and she was approaching us; and then I saw her sidelights. I saw her green light and the red. She was end-on to us. Later I saw her red light. This is all with the aid of a pair of glasses up to now. Afterwards I saw the ship's red light with my naked eye, and the two masthead lights. The only description of the ship that I could give is that she was, or I judged her to be, a four-masted steamer.'

Asquith: 'Why did you judge that?'

Boxhall: 'By the position of her masthead lights; they were close together.'

Asquith: 'Did the ship make any sort of answer, as far as you could see, to your rockets?'

Boxhall: 'I did not see it. Some people say she did, and others say she did not. There were a lot of men on the bridge. I had a quartermaster with me, and the captain [Smith] was standing by, at different times, watching this steamer.'

Asquith: 'Do you mean you heard someone say she was answering your signals?'

Boxhall: 'Yes, I did, and then she got close enough, and I Morsed to her – used our Morse [code] lamp.'

Asquith: 'You began Morsing [code] to her?'

Boxhall: 'Yes.'

Asquith: 'When people said to you that your signals were being answered, did they say how they were being answered?'

Boxhall: 'I think I heard somebody say that she showed a light.'

Asquith: 'Do you mean that she would be using a Morse [code] lamp?'

Boxhall: 'Quite probably.'

Asquith: 'Then you thought she was near enough to Morse [code] her from the "*Titanic*"?'

Boxhall: 'Yes, I do think so; I think so yes.'

Commissioner: 'What distance did you suppose her to be away?'

Boxhall: 'I judged her to be between 5 and 6 miles when I Morse [code] to her, and then she turned round – she was turning very, very slowly – until at last I only saw her stern light, and that was just before I went away in the [life]boat.'

Asquith: 'Did she make any sort of answer to your Morse [code] signals?'

Boxhall: 'I did not see any answer whatever.'

Asquith: 'Did anyone else, so far as you know, see an answer?'

Boxhall: 'Some people say they saw lights, but I did not.'

Asquith: 'Did they think they saw them Morsing in answer to your Morse [code] signals; did anyone say that?'

Boxhall: 'They did not say she Morse [code], but they said she showed a light. Then I got the quartermaster who was with me to call her up with our lamps, so that I could use the glasses to see if I could see signs of any answer; but I could not see any.'

Asquith: 'You could not see any with the glasses?'

Boxhall: 'No; and Captain Smith also looked, and he could not see any answer.'

Asquith: 'He also looked at her through the glasses?'

Boxhall: 'Yes.'

Asquith: 'After a time you saw what you took to be the stern light of a ship?'

Boxhall: 'It was the stern light of the ship.'

Asquith: 'Did you infer from that that the ship was turned round, and was going in the opposite direction?'

Boxhall: 'Yes.'

Asquith: 'When you first saw her, I understand you to say she was approaching you?'

Boxhall: 'She was approaching us, yes.'

Asquith: 'For about how long did you signal before it seemed to you that she turned round?'

Boxhall: 'I cannot say; I cannot judge any of the times at all.'

Asquith: 'Do you know at all whether the *"Titanic"* was swinging at this time?'

Boxhall: 'No, I do not see how it was possible for the *"Titanic"* to be swinging after the engines were stopped. I forget when it was I noticed the engines were stopped, but I did notice it; and there was absolutely nothing to cause the *"Titanic"* to swing.'

Asquith: 'After sending up those signals for some time did you turn your attention to the [life]boats?'

Boxhall: 'I was sending the rockets up right to the very last minute when I was sent away in the [life]boat.'

Asquith: 'When you say right up to the last minute, can you give me any idea of what you mean by that?'

Boxhall: 'Yes, right up to the time I was sent away in the [life]boat.'

Asquith: 'How long before the vessel sank were you sent away in the [life]boat?'

Boxhall: 'I cannot give the time, but I have approximated it nearly half an hour, as near as I could tell.'

Asquith: 'What [life]boat was it you were sent away in?'

Boxhall: 'In the emergency [life]boat number 2.'

Commissioner: 'It would be about a quarter to 2.'

Asquith: 'Yes, My Lord.'

Asquith: [To Boxhall] 'Who was superintending the filling of that [life]boat?'

Boxhall: 'Mr Wilde, or, I presume, Mr Wilde was superintending the filling. The order was given to lower away when I was told to go in it and the [life]boat was full; they had started the tackles when I got in.'

Commissioner: 'What number [lifeboat] was it?'

Boxhall: 'Port [lifeboat] number 2.'

Commissioner: 'Did you notice what other [life]boats there were on the port side at the time?'

Boxhall: 'There was only one [life]boat hanging there in the davits, number 4.'

Asquith: 'That was the [life]boat next to yours?'

Boxhall: 'Yes.'

Asquith: 'Can you say how many people were in that [life]boat number 2?'

Boxhall: 'I endeavoured to count them, but I did not succeed very well. I judge between 25 and 30 [people] were in her.'

Asquith: 'Were they mostly women, or were they mixed men and women?'

Boxhall: 'The majority were women. I know there were 3 crew, 1 male passenger, and myself.'

Asquith: 'And you think the rest were women?'

Boxhall: 'They were. There were several children in the [life]boat.'

Asquith: 'We have had evidence about this [life]boat from [James] Johnson, the steward, at page 91, and his evidence exactly corresponds with this. It is from question 3468 to about question 3478. He [Johnson] says he thinks there were 23 or 25 people in the [life]boat, and he afterwards says, "There was one male passenger and I think four members of the crew."'

Commissioner: 'This was an emergency [life]boat.'

Asquith: 'Yes, My Lord.'

Asquith: [To Boxhall] 'Did you notice when the people were being put in that [life]boat number 2 whether there were many passengers on deck at the time, round about?'

Boxhall: 'I did not notice the passengers being put into the [life]boat. I was not taking any notice of the boat at all, until I was sent to her.'

Asquith: 'Did you notice whether there were passengers on the deck at the time the boat was lowered?'

Boxhall: 'Yes, there were passengers round the deck, but I noticed as I was being lowered that they were filling number 4 [life]boat.'

Asquith: 'Were there any women about?'

Boxhall: 'I did not see any women.'

Asquith: 'I do not know whether you can say with regard to the starboard [life]boats at all whether there were any starboard [life]boats on the "*Titanic*" at this time, or whether they had all gone?

Boxhall: No, I cannot say. I know the starboard emergency [life]boat had gone some time, and that they were working on the collapsible [life]boats when I went, because I fired the distress signals from the socket in the rail just close to the bows of the emergency [life]boat on the starboard side. Every time I fired a signal I had to clear everybody away from the vicinity of this socket, and then I remember the last one or two distress signals I sent off the [life]boat had gone, and they were then working on the collapsible [life]boat which was on the deck.'

Asquith: 'Had you any lamp in your [life]boat number 2?'

Boxhall: 'Yes.'

Asquith: 'Had you put that in yourself or did you find it there?'

Boxhall: 'There is always a lamp in the emergency [life]boats.'

Asquith: 'Lamps are always kept there?'

Boxhall: 'They are lighted every night at 6 o'clock.'

Asquith: 'Do you mean they are not kept in the other [life]boats usually?'

Boxhall: 'They were not kept in the other [life]boats, no.'

Asquith: 'Did you see any put in the other [life]boats?'

Boxhall: 'Yes.'

Asquith: 'Was that by your orders?'

Boxhall: 'Well, it was through my speaking to the chief officer [Wilde] about it. I mentioned to him that there were no lamps. That was earlier on, when they started to clear the [life]boats. I mentioned to him the fact that there were no lamps in any of the boats, or compasses, and he told me to get hold of the lamp trimmer.'

Commissioner: 'When did you notice this?'

Boxhall: 'Oh, shortly after the orders were given to clear the [life]boats.'

Commissioner: 'You said "in any of the [life]boats." Did you examine all the [life]boats?'

Boxhall: 'Did I examine the [life]boats after the accident?'

Commissioner: 'Yes?'

Boxhall: 'No, I did not.'

Commissioner: 'Then you cannot speak from your knowledge?'

Boxhall: 'I examined the [life]boats on purpose. The lamps were in the lamp-room then.'

Commissioner: 'The lamps are in the lamp-room; the compasses are apparently kept in some locker; that is right, is it?'

Boxhall: 'Yes.'

Asquith: 'Did you have the lamps taken up?'

Boxhall: 'Yes. The chief officer [Wilde] told me to find the lamp trimmer. I did find him after a little trouble. I really forget where

I found him. He was on the boat deck working amongst the men. I told him to take a couple of men down with him and fetch the lamps, and he was afterwards seen to bring the lamps along the deck and put them in the [life]boats.'

Asquith: 'Do you know how many lamps were put into how many [life]boats?'

Boxhall: 'No, I do not know.'

Asquith: 'In your [life]boat did you also put in some green lights?'

Boxhall: 'Yes, there were some green lights lying in the wheelhouse. I told the quartermaster or someone who was around there to put them in the [life]boat.'

Asquith: 'Was any order given to you when you were lowered with regard to what you should do when you got into the water?'

Boxhall: 'No, I do not remember any.'

Asquith: 'What did you do when you got into the water?'

Boxhall: 'I pulled a little way from the ship, probably 100 feet away from the ship, and remained there for a while.'

Asquith: 'How long did you remain there; did you remain there until the ship sank?'

Boxhall: 'Oh, no, I did not. I did not remain there very long. I got the crew squared up and the oars out properly and the [life]boat squared when I heard somebody singing out from the ship, I do not know who it was, with a megaphone, for some of the boats to come back again, and to the best of my recollection they said "Come round the starboard side," so I pulled round the starboard side to the stern and had a little difficulty in getting round there.'

Asquith: 'Why was that, because you had not enough people to row?'

Boxhall: 'I had not enough people; my boat was rather deep. I had only one man who seemed to understand boat orders. I was pulling the stroke oar and trying to steer the boat at the same time myself.'

Asquith: 'There was only one seaman in your [life]boat?'

Boxhall: 'That is all.'

Asquith: 'Do you know whether there was a man named Osman?'

Boxhall: 'Yes, Osman or Osram, or something like that [Able Seaman Frank Osman (1885–1938)].'

Asquith: 'Who else rowed besides you and the seaman [Osman]? You were rowing and steering at the same time?'

Boxhall: Everybody was rowing with the exception of a male passenger. He did not seem to do much.

Asquith: 'You have told us there were two stewards or a steward and a scullaryman. They were both rowing?'

Boxhall: 'Oh, yes, they were rowing.'

Asquith: 'With some difficulty you rowed round to the starboard side of the ship?'

Boxhall: 'Yes, round the stern.'

Asquith: 'What did you do when you got round to the starboard side?'

Boxhall: 'Well, I stayed round on the starboard side, probably about 200 feet away from the ship. I found there was a little suction and I decided that it was very unwise to have gone back to the ship so I pulled away.'

Asquith: 'A little suction?'

Boxhall: 'Yes, there was a little suction.'

Asquith: 'Why was there suction at this time?'

Boxhall: 'The ship settling down badly, I suppose.'

Asquith: 'Was it settling down rapidly. Could you see it settling down at this time?'

Boxhall: 'Yes, I could see her settling down; I was watching the lines of lights.'

Commissioner: 'She was settling down by the head?'

Boxhall: 'She was settling down by the head, My Lord.'

Commissioner: 'Where were you at this time?'

Boxhall: 'Just a little, probably 200 feet, on the starboard beam of the ship, or probably a little abaft the starboard beam of the ship.'

Commissioner: 'Would there be any suction there?'

Boxhall: 'Well, I felt it; I saw it by the work we had pulling it round the ship's stern; seeing she was only a small boat, I judged there was quite a lot of suction.'

Asquith: 'Did you remain in that position, about 200 feet away from the ship, until she sank?'

Boxhall: 'No, I did not; I turned the [life]boat away and pulled in a north-easterly direction.'

Asquith: 'You mean, you pulled further away from the ship?'

Boxhall: 'Yes.'

Asquith: 'How far were you from the ship when she did sink?'

Boxhall: 'Approximately, half a mile.'

Asquith: 'That means that you could not see what happened?'

Boxhall: 'No, I could not.'

Asquith: 'After she sank, did you hear cries?'

Boxhall: 'Yes, I heard cries. I did not know when the lights went out that the ship had sunk. I saw the lights go out, but I did not know whether she had sunk or not, and then I heard the cries. I was showing green lights in the [life]boat then, to try and get the other [life]boats together, trying to keep us all together.'

Day Fourteen, 23 May 1912 (re-called)

Sir Robert Finlay: 'My Lord, I will not interrupt this witness (Turnbull) [George E. Turnbull] of course, but by-and-bye I desire permission to recall both Mr Lightoller and Mr Boxhall with regard to the question whether these messages were ever transmitted to the captain or any of the officers on board the *"Titanic"*?'

Commissioner: 'Very well, quite right.'

Sir Robert: 'As far as I recollect, no questions were put to Mr Lightoller on the subject at all or to Mr Boxhall, except in re-examination.'

Solicitor General: 'I think Sir Robert is right, and I realise, whatever we prove or do not prove with the help of the witness, we do not in any case by this evidence do more than carry messages to the Marconi office on the *"Titanic"*. What happened to them when they got into that office is not a matter which the present witness can tell at all.'

Commissioner: 'Nor does it matter.'

Solicitor General: 'Well, of course, it matters from the point of view of liability, because, of course, the Marconi operator may be regarded as not in the service of the White Star company; he may not be their servant, your Lordship sees.'

Commissioner: 'Well, in a sense, he is not.'

Sir Robert: 'I think, My Lord, in no sense is he in the service of the ship. Of course he is under discipline as everyone on board the ship must be, but he is not in their service; he is the servant of the Marconi Company.'

Commissioner: 'Still he is there, I suppose, for the very purpose, as a servant of the Marconi Company of communicating to the people in charge of the ship the messages which he gets which would affect the navigation.'

Sir Robert: 'Yes. I think we should be able without going into detail at the present moment to satisfy your Lordship beyond all doubt that these messages, the *"Mesaba"* message and the *"Amerika"* message were communicated either to the commander [Smith] or to any of the officers on board the *"Titanic"*.'

Fifth Officer Harold Godfrey Lowe

Harold Lowe was questioned in the United States on 23 April 1912, and by the British inquiry on days thirteen and fourteen as follows:

Day Thirteen, 22 May 1912

Mr Sidney Rowlatt: 'Harold Godfrey Lowe, is that your name?'

Lowe: 'Yes.'

Rowlatt: 'Were you the fifth officer on the *Titanic*?'

Lowe: 'I had that honour.'

Rowlatt: 'You have a master's certificate of competency?'

Lowe: 'I have.'

Rowlatt: 'I think you joined [*Titanic*] at Belfast, did you not?'

Lowe: 'I did.'

Rowlatt: 'Was it your duty to look at the [life]boats in Belfast and see that they were all there, and so on?'

Lowe: 'I was instructed by Mr Murdoch, the then chief officer [Wilde] of the ship, so do so.'

Rowlatt: 'Did you do it?'

Lowe: 'I did.'

Rowlatt: 'You went through the [life]boats and their equipment at Belfast?'

Lowe: 'Yes I, in the company of Mr Moody, I went.'

Rowlatt: 'He was lost?'

Lowe: 'We went through the starboard [life]boats.'

Rowlatt: 'Not the portside [life]boats?'

Lowe: 'Not the portside [life]boats?'

Rowlatt: 'Did anybody go through the portside [life]boats?'

Lowe: 'Mr Boxhall and Mr Pitman went through the portside [life]boats.'

Rowlatt: 'I will ask you in detail about that, but you sailed on the voyage. What was your watch on the Sunday of the accident?'

Lowe: 'My watch was the afternoon watch from 12 to 4 pm, and from 6 to 8 pm in the evening.'

Rowlatt: 'When did you go on after that?'

Lowe: 'At midnight.'

Rowlatt: 'You were on duty from 6 to 8 [pm]?'

Lowe: 'I was.'

Rowlatt: 'Did you hear anything about any messages about ice?'

Lowe: 'There was a chit on the chartroom table with the word "ice" on it.'

Rowlatt: 'You mean a little piece of paper with "ice" written on it?'

Lowe: 'A square chit of paper about 3 x 3.'

Rowlatt: 'On the chartroom table?'

Lowe: 'On the chartroom table.'

Rowlatt: 'What is that – "our chartroom table"?'

Lowe: 'The officers' chartroom table, and the word "ice" was written on top and then in a position underneath.'

Rowlatt: 'Can you remember what the position was?'

Lowe: 'I cannot.'

Rowlatt: 'Is that all that was brought to your attention about ice that day?'

Lowe: 'That is all.'

Rowlatt: 'Did you hear of Marconigrams coming about ice?'

Lowe: 'That was the only information I saw regarding ice.'

Rowlatt: 'That is all you have to say about your knowledge of ice on board the ship on that day?'

Lowe: 'Yes, that is all I know about it.'

Rowlatt: 'You went off watch at 8 o'clock?'

Lowe: 'Yes.'

Rowlatt: 'Did you turn in?'

Lowe: 'I went to bed.'

Rowlatt: 'Were you asleep at the time of the collision?'

Lowe: 'I was.'

Rowlatt: 'Just tell us what woke you up?'

Lowe: 'I was half awakened by hearing voices in our quarters, because it is an unusual thing, and it woke me up. I suppose I lay there for a little while until I fully realised, and then I jumped out of bed and opened my door a bit and looked out, and I saw ladies in our quarters with lifebelts on.'

Rowlatt: 'When you first looked out people had got their lifebelts on?'

Lowe: 'They had.'

Rowlatt: 'Do you know the time?'

Lowe: 'I do not. I have not the remotest idea of the time right throughout.'

Rowlatt: 'Were the [life]boats being attended to?'

Lowe: 'As soon as I looked out through the door I jumped back and got dressed and went out on deck and the [life]boats were being cleared.'

Commissioner: 'The [life]boats had been cleared did you say?'

Lowe: 'The [life]boats were being cleared.'

Rowlatt: 'Did you go to the starboard side first?'

Lowe: 'I had to go round to the portside first, that is on my way to the starboard.'

Rowlatt: 'As you were round the portside, the [life]boats there were being cleared, were they?'

Lowe: 'Yes.'

Rowlatt: 'Did you take any part in clearing the [life]boats or have anything to do in connection with them on the portside?'

Lowe: 'No.'

Rowlatt: 'You got to the starboard side?'

Lowe: 'I got to the starboard side.'

Rowlatt: 'What [life]boat did you get to?'

Lowe: 'The first [life]boat I went to was number 7.'

Rowlatt: 'That would be the aftermost one upon the starboard side?'

Lowe: 'No. That would be the afterboat of the forward section.'

Rowlatt: 'You came round behind the deck house?'

Lowe: 'I came round abaft the second funnel.'

Rowlatt: 'Was that the [life]boat to which you belonged?'

Lowe: 'No.'

Rowlatt: 'What was the [life]boat to which you belonged?'

Lowe: 'I do not know.'

Commissioner: 'Why do you not know?'

Lowe: 'I do not know why, but I do not.'

Commissioner: 'Was it your business to find out?'

Lowe: 'I suppose it was.'

Commissioner: 'And you did not do it?'

Lowe: 'No, sir.'

Rowlatt: 'Why did you go round to [lifeboat] number 7?'

Lowe: 'Because the people were there.'

Rowlatt: 'What was being done at [lifeboat] number 7?'

Lowe: 'Loading it [the lifeboat] with women and children.'

Rowlatt: 'Did you assist there?'

Lowe: 'I did.'

Rowlatt: 'Did you see that [life]boat lowered?'

Lowe: 'I did. I assisted in lowering it.'

Rowlatt: 'Then did you go to [lifeboat] number 5?'

Lowe: 'I went to [lifeboat] number 5.'

Rowlatt: 'Did you see that [lifeboat] lowered?'

Lowe: 'I did.'

Rowlatt: 'Did you assist?'

Lowe: 'I did.'

Rowlatt: 'When you say you assisted, did you take charge of the operations?'

Lowe: 'I assisted; that is to say, Mr Murdoch was superintending.'

Rowlatt: 'Mr Murdoch was there?'

Lowe: 'Yes.'

Rowlatt: 'Then was [lifeboat] number 5 lowered after number 7?'

Lowe: '[Lifeboat] number 5 was lowered after number 7.'

Rowlatt: 'Did you then go to [lifeboat] number 3?'

Lowe: 'I then went to [lifeboat] number 3.'

Rowlatt: 'Was that lowered?'

Lowe: 'That was lowered.'

Rowlatt: 'And did you then go to the emergency [life]boat?'

Lowe: 'I went to number 1, the emergency [life]boat.'

Rowlatt: 'Was that lowered?'

Lowe: 'Yes.'

Rowlatt: 'When your [life]boat was lowered that lot of [life]boats were finished with. Did you notice any list?'

Lowe: 'No.'

Rowlatt: 'Was the vessel down by the head?'

Lowe: 'Yes.'

Rowlatt: 'You noticed that?'

Lowe: 'Yes, of course I noticed that as soon as I got up.'

Rowlatt: 'Did you look for any lights at this time at all?'

Lowe: 'I was getting the emergency [life]boat ready, number 1, Mr Boxhall was firing the detonators, the distress signals, and somebody mentioned something about a ship in the port bow, and I glanced over in that direction casually and I saw a steamer there.'

Rowlatt: 'What did you see of her?'

Lowe: 'I saw her two mastheads and her red side-lights.'

Rowlatt: 'That accounts for all these four [life]boats?'

Lowe: 'Yes, the forward section.'

Rowlatt: 'Where did you go then?'

Lowe: 'I then went to [lifeboat] number 14.'

Rowlatt: 'That is right aft on the other side, is it not?'

Lowe: 'That would be the second forward [life]boat of the after section, and the second [life]boat from aft of the after section.'

Rowlatt: 'Why did you go to her in particular?'

Lowe: 'Because they seemed to be busy there.'

Rowlatt: 'Did you got assist there?'

Lowe: 'I did.'

Rowlatt: 'Who was in charge there?'

Lowe: 'I do not know who was in charge there. I finished up loading [lifeboat] number 14 and Mr Moody was finishing up loading [lifeboat] number 16.'

Rowlatt: 'Did you see anything about [lifeboat] number 12?'

Lowe: '[Lifeboat] number 12 would be the forward [life]boat – the boat next to me forward? – Yes. [Lifeboat] numbers 12, 14 and 16 went down pretty much at the same time.'

Rowlatt: 'You went in number [lifeboat] 14, did you not?'

Lowe: 'Yes.'

Rowlett: 'Did you go by anybody's orders?'

Lowe: 'I did not. I saw five boats go away without an officer, and I told Mr Moody on my own that I had seen five [life]boats go away, and an officer ought to go in one of these [life]boats. I asked him who it was to be – him or I – and he told me, "You go, I will get in another [life]boat."'

Rowlett: 'Were you lowered in that [life]boat?'

Lowe: 'I was lowered in [lifeboat] number 14.'

Rowlatt: 'I want to ask you a little about that. Was there any difficulty in lowering when you got near the water?'

Lowe: 'Yes, I slipped her.'

Rowlatt: 'Did the falls go wrong?'

Lowe: 'Something got wrong and I slipped her.'

Rowlatt: 'That means to say, you threw off the lever when you were some way from the water?'

Lowe: 'I should say I dropped her about five feet.'

Rowlett: 'Your Lordship remembers that Scarrott told us about that. Was that because the falls...?'

Lowe: 'That was because I was not going to wait and chance being dipped down by the stem by anybody on top, so I thought it was best for me to drop and know what I was doing.'

Rowlatt: 'No doubt you dealt with the situation quite rightly, but I want to know what caused the situation. Was it because the rope would not run any further?'

Lowe: 'I do not know, because, you must understand that the lowering away was being carried out on deck, and I must have been about 64 feet below that deck, and I could not see it.'

Rowlatt: 'Did you look up?'

Lowe: 'Yes.'

Rowlatt: 'Could you tell me why you were not being lowered further?'

Lowe: 'No.'

Rowlatt: 'You could not?'

Lowe: 'No.'

Rowlatt: 'One of the men in your [life]boat has given evidence, and he says he looked up and saw the rope of the falls twisted?'

Lowe: 'No; I looked up and I could not see anything.'

Rowlatt: 'Just let me ask you this, because it is fair to ask you it. Could they twist?'

Lowe: 'I suppose they could.'

Rowlatt: 'Can the blocks revolve at the top?'

Lowe: 'Oh, yes, the blocks are movable in the davits; they are swivelled; both are swivelled, the top and bottom blocks.'

Rowlatt: 'Then you got to the water and you slipped her, as you say?'

Lowe: 'Yes.'

Rowlatt: 'Did you take command of the [life]boat?'

Lowe: 'Yes.'

Rowlatt: 'What did you do with her?'

Lowe: 'I took, I think it was, [lifeboat] number 12 to a distance of about 150 yards from the ship, and told him to stay there until I gave him orders to go away or any other orders. I then came back to the ship and escorted another [life]boat, and so on, until I had five [life]boats there.'

Rowlatt: 'You gathered five [life]boats together?'

Lowe: 'Yes.'

Rowlatt: 'There is just another thing I want to ask you. Did you use a revolver at all?'

Lowe: 'I did.'

Rowlatt: 'How was that?'

Lowe: 'It was because while I was on the boat deck just as they had started to lower, two men jumped into my [life]boat. I chased one out and to avoid another occurrence of that sort I fired my revolver as I was going down each deck, because the [life]boat would not stand a sudden jerk. She was loaded already I suppose with about 64 people on her, and she would not stand any more.'

Rowlatt: 'You were afraid of the effect of any person jumping in the [life]boat through the air?'

Lowe: 'Certainly, I was.'

Rowlatt: 'In your judgment had she enough in her to lower safely?'

Lowe: 'She had too many in her as far as that goes. I was taking risks.'

Rowlatt: 'You say you collected these four [life]boats together at a distance of about 150 yards?'

Lowe: 'Yes.'

Rowlatt: 'Can you judge how long that was before the ship went down?'

Lowe: 'I have not the remotest idea of time from the time she went down until we boarded the *Carpathia*. All I know is that when we boarded the *Carpathia* in the morning it was six o'clock, and that is the only time I know of.'

Rowlatt: 'You could not give me any idea?'

Lowe: 'I could not; it is no good my trying.'

Rowlatt: 'What did you do after you got the four [life]boats out there?'

Lowe: 'I tied them together in a string, and made them step their masts.'

Rowlatt: 'What was that for?'

Lowe: 'In case it came on to blow, and then they would be ready.'

Rowlatt: 'Did you transfer any of your passengers?'

Lowe: 'Yes, I transferred all of them.'

Rowlatt: 'Among the other [life]boats?'

Lowe: 'Into the other four [life]boats.'

Rowlatt: 'Why did you do that?'

Lowe: 'So as to have an empty [life]boat to go back.'

Commissioner: 'To do what?'

Lowe: 'To go back to the wreck.'

Rowlatt: 'Was that before the *Titanic* foundered or after?'

Lowe: 'No, that was after she went down.'

Rowlatt: 'Having got an empty [life]boat, did you go back to the wreckage?'

Lowe: 'I did.'

Rowlatt: 'Was there much wreckage?'

Lowe: 'No, very little.'

Commissioner: 'Am I to understand that you were alone in the [life]boat?'

Lowe: 'No.'

Rowlatt: 'You were there with your crew?'

Lowe: 'Yes.'

Commissioner: 'How many men had you in the [life]boat?'

Lowe: 'I do not know; I should say seven.'

Commissioner: 'Including yourself?'

Lowe: 'Yes, I should say six and myself.'

Rowlatt: 'Did you row six oars back to the wreck?'

Lowe: 'No, five oars, I think, and I had a man on the look-out.'

Rowlatt: 'I understand what you say is that you got rid of the passengers. You got rid of the people who could not do anything, and went back with a working crew to look for people who were drowning; that is what you mean?'

Lowe: 'Yes; it would be no good me going back with a load of people.'

Rowlatt: 'Certainly; I am not complaining; I am only trying to bring it out in your favour, if I may say so. You rescued some people, did not you?'

Lowe: 'I picked up four.'

Rowlatt: 'I think one died in the [life]boat, did he not?'

Lowe: 'One died, a Mr Hoyt, of New York [William F. Hoyt (1869–1912)]'.

Rowlatt: 'Were they men?'

Lowe: 'Four men.'

Rowlatt: 'Did you see any other people alive?'

Lowe: 'Not one, or else I should have picked them up.'

Rowlatt: 'Did you see bodies?'

Lowe: 'Yes.'

Rowlatt: 'After that did you come across the submerged collapsible [lifeboat] of which we have heard?'

Lowe: 'Yes.'

Rowlatt: 'It was you who took the people off that, was it?'

Lowe: 'I did.'

Rowlatt: 'Was it the one with Mr Lightoller on board?'

Lowe: 'No, it was not.'

Rowlatt: 'Another one?'

Lowe: 'Another one.'

Rowlatt: 'Were there two submerged collapsibles [lifeboats]?'

Lowe: 'I do not know – I did not know at the time, but, of course, I know now. The one that I picked up, I reckon, had been pierced, but I do not know. She was right side up and all that.'

Rowlatt: 'Was she extended, or whatever you call it, opened out; were the collapsible sides pulled up?'

Lowe: 'No, the sides had dropped somehow or other.'

Rowlatt: 'She was flat?'

Lowe: 'She was right side up.'

Rowlatt: 'Can you give us any idea of who were on board of her – you do not know?'

Lowe: 'No. I can only give you one, and that was the lady that was on board there.'

Rowlatt: 'The lady?'

Lowe: 'Yes.'

Rowlatt: 'Can you tell me how many collapsibles [lifeboats] got to the *Carpathia*, because we cannot account for the collapsibles [lifeboats]?'

Lowe: 'I abandoned one, and then I towed another one while I was under sail to the *Carpathia*; that is two; then the one that Mr Lightoller was on, that is three. I do not know where the fourth is.'

Rowlatt: 'So far as you know there were only three ever got away from the wreck in any shape?'

Lowe: 'As far as I know.'

Thomas Scanlan MP: 'You stated in giving evidence in America that a crowd went down to the gangway doors to get them open, and that you were going to load the [life]boats and take passengers in from these gangway doors?'

Lowe: 'I did.'

Scanlan: 'It has come out in the evidence that a number of women and children perished on the *Titanic*. I believe that is a fact. May it be that in the expectation of this method being carried out, a number of the women and children were directed down to these gangways?'

Lowe: 'No, it is not.'

Scanlan: 'Were you giving directions as to the filling of [life]boat number 1?'

Lowe: 'I was.'

Scanlan: 'And the lowering of her?'

Lowe: 'And the lowering of her.'

Scanlan: 'She was loaded with a very small number of passengers – five?'

Lowe: 'I do not know how many there were. I took everybody that was there; that is all I know.'

Commissioner: 'You took what?'

Lowe: 'I cleared the deck, My Lord.'

Commissioner: 'You mean to say that when you took the people into [lifeboat] number 1 there were no people left on the deck?'

Lowe: 'There were no people left on the starboard deck.'

Scanlan: 'At that time what search did you have made for people – for passengers?'

Lowe: 'I did not make any search.'

Scanlan: 'You did not, for instance, send over to the port side to find if there were any women or children?'

Lowe: 'No, because I wanted to get the [life]boats away. I did not have any time to waste.'

Scanlan: 'And you did not send down to any of the lower decks?'

Lowe: 'There was nobody on the next deck. I stopped the [life]boat there and asked them to look.'

Scanlan: 'Or on any of the lower decks?'

Lowe: 'I do not know about that. I stopped the lowering of the [life] boat at A Deck, and told the men to have a look there, and they saw nobody.'

Scanlan: 'There was no particular reason why that [life]boat should have been lowered with only five passengers?'

Lowe: 'No particular reason why the [life]boat should be lowered with only five people.'

Commissioner: 'You are following a bad example, Mr Scanlan. Instead of asking questions, you are making a statement, and I do not think your statement is in accordance with his [Lowe's] evidence.'

Scanlan: 'I appreciate the mistake, My Lord.'

Scanlan: [To Lowe] 'At the time that [life]boat number 1 was lowered there were still other [life]boats on the starboard side?'

Lowe: 'That I am not prepared to answer; I do not know.'

Scanlan: 'I mean [life]boats were lowered after [lifeboat] number 1?'

Lowe: 'I say I do not know.'

Day Fourteen, 23 May 1912 (re-called)

Sir Robert Finlay: 'You were on duty from 6 to 8 [pm]?'

Lowe: 'I was.'

Sir Robert: 'Did you ever hear anything about a message from the "*Amerika*" to be sent on to Cape Race about ice?'

Lowe: 'No.'

Sir Robert: 'Or about a message from the "*Mesaba*"?'

Lowe: 'No.'

Solicitor General: 'You have told us before about the chit that you saw, the little piece of paper on the chart room table. Can you tell us when you saw it?'

Lowe: 'I suppose it must have been shortly after 6 [pm].'

Solicitor General: 'Shortly after 6 [pm]?'

Lowe: 'Yes.'

Solicitor General: 'What had it on it actually?'

Lowe: 'It had the word "ice", and then a position underneath the word "ice".'

Solicitor General: 'What do you mean by a "position"?'

Lowe: 'That means to say a latitude and a longitude.'

Commissioner: 'Where the ice was?'

Lowe: 'Where the ice was, My Lord, yes.'

Solicitor General: 'Did you recognise the writing?'

Lowe: 'I did not. I suppose I only looked at it casually.'

Solicitor General: 'I think you told us before that when you saw it, you reckoned on the chart when you would get to it?'

Lowe: 'I did not say when we would meet it; I said that I worked it out mentally, and that I found that we should not come to that position during my watch from 6 to 8 [pm]. That is what I meant to imply.'

Solicitor General: 'I think that is exactly what you said. You are quite right. You cannot tell us more about it than that?'

Lowe: 'I cannot.'

Solicitor General: 'I will put the same question to you, and that is the only other thing I want to know. How many reports about ice – I am not talking about tank steamers – did you hear of?'

Lowe: 'I do not remember having heard of any, and that is the only one that I saw. They may have been on the notice board, and I may not have looked at the notice board. I do not remember looking at the notice board, and that is the only paper or note that I saw referring to ice, as I have stated.'

BIBLIOGRAPHY AND RESEARCH SOURCES

A & E Television Networks, Titanic: *The Complete Story* (Television Documentary), 1994

Ancestry.co.uk

A Night to Remember (Film), 1958

Atlantic Daily Bulletin, The Journal of the British *Titanic* Society

Auckland Evening Post, 30 June 1922

Australian National Maritime Museum

Babler, Gunter, *Guide to the Crew of* Titanic: *The Structure of Working aboard the Legendary Liner*, 2017

Ballard, Robert Duane, *The Discovery of the* Titanic, 1987

Bancroft, James W., *The* Titanic *Disaster: Omens, Mysteries and Misfortunes of the Doomed Liner*, Pen & Sword Books, 2023

Bancroft, James W., Titanic: *Iceberg Ahead – The Story of the Disaster by Some of Those Who Were There*, Pen & Sword Books, 2021

Beed, Blair Stephen, Titanic *Victims in Halifax Graveyards*, 2001

Beesley, Lawrence, *the Loss of the SS* Titanic: *Its Story and Its Lessons*, 1912

Behe, George, *On Board RMS* Titanic: *Memories of the Maiden Voyage*, 2012

Behe, George, *The Triumvirate: Captain Edward J. Smith, Bruce Ismay, Thomas Andrews and the Sinking of* Titanic, 2024

Belfast *Titanic* Society

Bell, Chief Engineer Joseph: Letter to his son written on 11 April 1912

Beveridge, Bruce and Hall, Steve, *RMS* Titanic *in 50 Objects: With Images of Artefacts from the Collection of White Star Memories*, 2022

Board of Trade, *The Investigation into the Loss of the SS* Titanic, 1912

Boxhall, Commander Joseph, Series of letters written to Joseph Carvalho in Massachusetts in 1961 and 1962

Boxhall, Joseph Groves, His diary dated from 1899 to 1902

Boxhall, Commander Joseph, Transcript of BBC Radio broadcast on 22 October 1962

Bridgwater Mercury, 12 April 1912

Brisbane Telegraph, 4 June 1912

British Government: *Loss of the Steamship* Titanic: *Report of a Formal Investigation into the Circumstances Attending the Foundering on April 15, 1912 of the British Steamship Titanic, of Liverpool, After Striking Ice in or near Latitude 41 – 46 N; Longitude 50 – 14 W, North Atlantic Ocean, as Conducted by the British Government*, 1912
British *Titanic* Society
Bullock, Stan F., *A Titanic Hero: Thomas Andrews, Shipbuilder*, 1912

Cameron, Stephen: Titanic: *Belfast's Own*, 1998
Cambrian News and Merionethshire Standard, 4 July 1919
Census Returns, 1841–1921
Chicago Daily Journal, 19 April 1912 and 16 January 1914
Chirnside, Mark, *The 'Olympic' Class Ships:* Olympic, Titanic *and* Britannic, 2011
Chirnside, Mark, *The Sting of the* Hawke: *Collision in the Solent – The Full Story Behind the Collision between HMS* Hawke *and RMS* Olympic *on 20 September 1911*, 2015
Chirnside, Mark, *The Big Four of the White Star Fleet:* Celtic, Cedric, Baltic *and* Adriatic, 2016
Chirnside, Mark, *Oceanic: White Star's Ship of the Century*, 2018
Colley, Edward, Letter to his sister-in-law in 1912
Cooper, G.J., Titanic *Captain: The Life of Edward John Smith*, 2011

Daily Herald, 23 May 1912
Daily News (Perth), 7 April 1933
Daily Sketch, 25 April 1912 and 30 April 1912
Daly, Eugene, Letter to his sister, 1912
Davenport-Hines, Richard: Titanic *Lives: Migrants and Millionaires, Conmen and Crew*, 2012
Dumfries and Galloway Standard and Advertiser, 21 August 1912

Edkins, Richard, *Murdoch of the* Titanic
Encyclopaedia Titanica
Everett, Marshall: *Wreck and Sinking of the* Titanic, *the Oceans Greatest Disaster*, 1912

FindMyPast
Fox, Stephen R., *Transatlantic: Samuel Cunard, Isambard Brunel, and the Great Atlantic Steamships*, 2003

Geller, Judith B., Titanic: *Women and Children First*, 1998
Gibbons, Elizabeth, *To the Bitter End*, 1992
Gibbs, Phillip, *The Deathless Story of the* Titanic: *Lloyds Weekly News*, 1912
Grace's Guide to British Industrial History
Gracie, Archibald, *The Truth about the* Titanic, 1973
Grimsby Daily Telegraph, 4 April 1992

Halpern, Samuel, *Report into the Loss of the SS* Titanic: *A Centennial Appraisal*, 2011
Harland and Wolff Director's Minute Book held in the Public Records Office
Harris, Rene, *Her Husband Went Down With The* Titanic. *Liberty Magazine* for 23 April 1932

Hart's Army Lists
Holman, Hannah, Titanic *Voices: 63 Survivors Tell Their Extraordinary Stories*, 2011
Honolulu Star-Bulletin, 3 August 1912
Hull Daily Mail for 27 March 1905, 16 April 1912, 5 October 1914, 27 March 1919, 10 September 1929
Hume, Yvonne and Hume, John Law, and Dean, Millvina: *RMS* Titanic: *The First Violin; the True Story of* Titanic's *First Violinist*
Hyslop, Donald; Forsyth, Alastair; Jemima, Sheila: Titanic *Voices: Memories from the Fateful Voyage*, 1994

Illustrated London News, 4 May 1912
Imperial War Museum, London
Ismay, Clifford, *Understanding J. Bruce Ismay: The True Story of the Man they Called 'The Coward of the* Titanic*'*, 2022

JWB Historical Archive

Lancashire Online Parish Clerk Records
Life (Periodical)
Lightoller, Charles Herbert, Titanic *and Other Ships*, 1935
Liverpool Journal of Commerce, 21 May 1912
Liverpool Record Office
Lives of the First World War, 1914–1918
Lloyds List
London Daily News
Lord, Walter, *A Night to Remember*, 1955
Lord, Walter, *The Night Lives On*: *Thoughts, Theories and Revelations about the* Titanic, 1986
Lowell Sun (Massachusetts), 28 May 1912

McDonald, Ronnie, *Charles Lightoller Secrets, St Margaret's Community Website*, 2010
Maidenhead Advertiser, 29 April 1912
Marcus, Geoffrey Jules, *The Maiden Voyage*, 1969
Merseyside Maritime Museum
Moss, Michael S. and Hume, John R., *Shipbuilders to the World: 125 Years of Harland and Wolf, Belfast, 1861–1986*, 1986
Myers, L.T., *The Sinking of the* Titanic *and Great Sea Disasters*, 1912

National Geographic (Periodical)
National Maritime Museum
National Museums, Liverpool, Titanic *and Liverpool: The Untold Story*, 2012
Nautical Magazine, May 1959
New York World, 21 April 1912
Nicholas, Anthony, *Key Figures aboard RMS* Titanic: *Superstars and Scapegoats*, 2022
North Wales Weekly News, 26 September 1913

Oldham, Wilton J., *The Ismay Line: The White Star Line and the Ismay Family Story*, 1961

Padfield, Peter, *The* Titanic *and the* Californian, 1965
Patten, Louise, *As Good As Gold*, 2010
Pellow, James with Kendle, Dorothy, A *Lifetime on the* Titanic: *The Biography of Edith Haisman, Britain's Oldest Survivor of the* Titanic *Disaster*, 1995
Pierce, Nicola, Titanic: *True Stories of Her Passengers, Crew and Legacy*, 2018
Piouffre, Gérard, *Le* Titanic *ne répond plus*, 2009

Rheims, George Alexander Lucien, Letter to his wife written in French on 19 April 1912
Rostron, Sir Arthur Henry, *Home from the Sea*, 1931
Royal Museums, Greenwich
Royal Naval Reserve Records, 1899–1930
Russell, Gareth, *The Ship of Dreams: The Sinking of the* Titanic *and the End of the Edwardian Era*, 2019 *Scotland on Sunday*, 8 March 1998

Sheffield Daily Telegraph, 7 August 1913
Sheil, Inger: Titanic *Valour: The Life of Fifth Officer Harold Lowe*, 2011
Sheil, Inger, *William McMaster Murdoch: A Career at Sea*, 2002
Sheil, Inger and Hyder, Gemma, *On Watch: The Deck Officers and Wireless Operators of the RMS* Titanic, 2002
Shiel, Inger and Sundberg, Kerri, *Bridge Duty Officers of the RMS* Titanic, 1999
Shepton Mallet Journal, 23 March 1956
Society of Naval Architects and Marine Engineers: Titanic: *The Anatomy of a Disaster: A Report from the Marine Forensic Panel*, 1997
Southampton Pictorial (various)
Southampton Times and Hampshire Express, 18 May 1912
Sphere, 25 May 1912
Stenson, Patrick: Titanic *Voyage: The Odyssey of C.H. Lightoller*, 1998
Stormer, Susanne, *Good-Bye, Good Luck: The Biography of William McMaster Murdoch*, 1995
Stringer, Craig, Titanic *People*, 2012

Taunton Courier and Western Advertiser, 2 August 1958
The Adelaide Observer, 9 November 1912
The Baltimore Enquirer, 20 July 1912
The Boston Globe, 23 April 1912
The Canberra Times, 18 October 1962 and 6 October 1985
The Central Somerset Gazette, 27 August 1954
The Chicago Daily Tribune, 20 April 1912
The Cincinnati Enquirer, 21 July 1912
The Daily Telegraph, 1 June 1911
The London Gazette, 5 March 1912, 25 June 1915 and 12 June 1923
The Manchester Guardian, July 1932
The Melbourne Argus, 22 April 1912
The National Archives
The New York Herald, 21 April 1912
The New York Times, 19 April 1912, 21 April 1912 and 8 July 1912

The Sydney Daily Telegraph, 18 June 1912
The World's News, 27 September 1924
The Yorkshire Post, 7 May 2003
Ticehurst, Brian, *The Titanic's Rescuers: Captain, Sir Arthur Rostron and the Crew of the Carpathia*, 1996
Titanic *Commutator*: The Journal of *Titanic* Historical Society
Titanic Historical Society
Titanic International Society
*Titanic*Officers.com
Turner, Steve, *the Band That Played On: The Extraordinary Story of the 8 Musicians Who Went Down With the Titanic*, 2011

United States Enquiry: *Investigation into the Loss of the SS Titanic*, 1912

Villiers, Alan, *Of Ships and Men: A Personal Anthology* 1962

Wade, Wyn Craig, *The Titanic: End of a Dream*, 1979
Wade, Wyn Craig, *The Titanic: Disaster of a Century*, 2012
Ward, Christopher, *And the Band Played On: The Titanic Violinist and the Glovemaker – A True Story of Love, Loss and Betrayal*, 2011
Welshman, Dr John, Titanic: *The Last Night of a Small Town*, 2012
Whiteley, Thomas, Titanic *Lecture*
Wilson, Andrew, *Shadow of the Titanic: The Extraordinary Stories of Those Who Survived*, 2012
Winocour, Jack, *The Story of the Titanic, as told by its Survivors*, 1960
Woods, Captain E.A., *The White Star Sailing Packets: Historic Society of Lancashire and Cheshire*, 1944
Wormstedt, Bill, Fitch, Tad and Layton, J. Kent, *On a Sea of Glass: The Life and Loss of RMS* Titanic, 2015

INDEX

Baltimore, USA, 81 84 85 87
Barmouth, Wales, 74 75
Belfast, 5 8 205
Births, 55 57 60 69 71 73 77
Blair, Second Officer David, x 6 10
Boxhall, Fourth Officer Joseph Groves, 7 8 11 18 23 25 27 28 30 35 38 39 44 60 71–73 156 174 181–205 211

Californian, RMS, 139 140 141 144 175
Carpathia, RMS, 36 43–44 46 82 218
Cherbourg, France, 3 14 17 93
Chorley, Lancashire, 6 60–61
Christchurch, Hampshire, 73

Dalbeattie, Scotland, 57
Deganwy, Conwy, Wales, 76
Deaths and burials, 57 60 68 71 73
Descriptions, 6 7 9 33

Eglwys Rhos, Llanrhos, Wales, 73

Haddock, Captain Herbert James, 5 7–8 56
Harland and Wolff, 2
Hawke, HMS, 3–4 5 10
Hull, Kingston-upon-, 7 71 73

Inquiries (British and American), 43 46 59 63 70 79–80 89–222
Ireland, 5 8 14 17
Ismay, Joseph Bruce, 1–2 15–16 28–29 46 52 83 159

Lifeboat 5, 8 14–15 29 157–158 165–166
Lifeboat 14, 8 14–15 40–41 212
Lifeboat Emergency 2, 8 14–15
Lifeboat Collapsible B, 8 14–15

Lightoller, Second Officer Charles Herbert, ix 6 8 10 11 23 25 30–31 32 34 35 36–37 38 43 51–52 57 58 59 60–69 91–147 151 172 184 191 218
Liverpool, 1 5 55–56 58–59 61 72 78 87–91
Lowe, Fifth Officer Harold Godfrey, ix 7 8 11 15 23 25 28 29 30 32 33 34 37–38 40–43 60 73–77 82 87 174 205–222

Marconigram, 106–108 134–135 173 177–180
Marriages, 56 58 63 70 72
Memorials, 57 60 68 71 73 78
Moody, Sixth Officer James Paul, 7 8 11 23 32 34 37–38 77–78 106 107 134 139
Murdoch, First Officer William McMaster, 3 5–6 8 11 17 23 27 28 29 35 37–38 57–60 81 94 95 98 132 136 155 161–162 169 172 186 187 189 205

New York, 1 3 44 51 56 89

Oceanic, RMS, 6 7 63 65
Olympic, RMS, x 3 5 10 56 70 73

Passengers, 14–16 17 18–19 21–22 28 32 35 36 41 42 45–53 81
Pitcombe, Somerset, 71
Pitman, Third Officer Herbert John, 6 8 11 17 23 25 28 38 39–40 44 60 69–71 148–181 191
Pryal, Captain Peter, 6 81 83–86

Royal Navy (and Reserve), 5 6 7 9 56 58 67 70 72 75 77

229

Scarborough, Yorkshire, 7 77 78
Shipwrecks, ix 6 45 61–62
Smith, Captain Edward John, x 3 6 8 9 10 11 22 23 27 28 31 32 45 46 50 79–87 95 97 98 107 120 121 131 132 133 146 159 169 187 189 193 196
Southampton, x 5 10 11 14 66 93 105
Sutton Montis, Somerset, 69

Titanic, RMS, ix x 3 4 5 8 10 11 45
Twickenham, 68

Washington (DC), USA, 44 87
White Star Line, 1–2 4 5 7 45 56 58 63 70 75 76 78 84 148 181
Wilde, Chief Officer Henry Tingle, x 5 6 9–10 11 17 30 32 35 37–38 55–57 93 94 103 109 112 132 172 197 200 205
World War One, 65–66 70 72 75
World War Two, 67–68 71 77